MARINE
NATURAL PRODUCTS

Chemical and Biological Perspectives

Volume IV

Contributors

MICHEL BARBIER

L. CHEVOLOT

TATSUO HIGA

RICHARD E. MOORE

PAUL J. SCHEUER

MARINE
NATURAL PRODUCTS

Chemical and Biological Perspectives

Volume IV

EDITED BY

PAUL J. SCHEUER

Department of Chemistry
University of Hawaii
Honolulu, Hawaii

ACADEMIC PRESS

A Subsidiary of Harcourt Brace Jovanovich, Publishers

London   New York   Toronto   Sydney   San Francisco   1981

ACADEMIC PRESS, INC.
111 Fifth Avenue, New York, New York 10003

United Kingdom Edition published by
ACADEMIC PRESS, INC. (LONDON) LTD.
24/28 Oval Road, London NW1 7DX

Library of Congress Cataloging in Publication Data

Main entry under title:

Marine natural products.

Includes bibliographies and indexes.
1. Natural products--Addresses, essays, lectures.
2. Biological chemistry--Addresses, essays, lectures.
3. Marine pharmacology--Addresses, essays, lectures.
I. Scheuer, Paul J.   II. Darias, J.
QD415.M28   547.7   77-10960
ISBN 0-12-624004-3 (v. 4)

PRINTED IN THE UNITED STATES OF AMERICA

81 82 83 84   9 8 7 6 5 4 3 2 1

Contents

Chapter 3   Phenolic Substances

TATSUO HIGA

Chapter 4   Marine Chemical Ecology: The Roles of Chemical Communication and Chemical Pollution

MICHEL BARBIER

Appendix   Register of Known Compounds

PAUL J. SCHEUER

Index

Contributors

Numbers in parentheses indicate the pages on which the authors' contributions begin.

MICHEL BARBIER (147), Institut de Chimie des Substances Naturelles, 91190 Gif-sur-Yvette, France

L. CHEVOLOT (53), Centre Océanologique de Bretagne, 29273 Brest Cedex, France

TATSUO HIGA (93), Department of Marine Sciences, University of Ryukyus, Nakagusuku, Okinawa, 901-24, Japan

RICHARD E. MOORE (1), Department of Chemistry, University of Hawaii at Manoa, Honolulu, Hawaii 96822

PAUL J. SCHEUER (187), Department of Chemistry, University of Hawaii at Manoa, Honolulu, Hawaii 96822

General Preface

"Chemistry of Marine Natural Products" (Academic Press, 1973), the progenitor of the present series, covered the early literature of a budding research area through December 1971. Since then, the field of marine natural products has flowered beyond expectation. Research has grown geometrically; it has spread geographically; and it has begun to explore in earnest some fascinating phenomena at the interface between biology and chemistry. Since March 1973, when "Chemistry of Marine Natural Products" was published, it has become increasingly apparent to me that a review of the entire field by one person was no longer feasible; hence the present effort in which I have asked some of my colleagues to share the task of providing critical reviews and new perspectives for the marine research community. I am grateful for the enthusiastic and prompt response by the contributors to this as well as to subsequent volumes.

Another facet of the 1973 book also needed reexamination. When I planned and wrote "Chemistry of Marine Natural Products," the organizational choices were essentially between a phyletic and a biological approach. I chose a broad structural biogenetic outline, a concept with which I was comfortable and which, in my opinion, filled a need at that time. Such a unidimensional design no longer seems satisfactory. It has now become desirable to highlight and review topics even though they may bear little lateral relationship to one another. It may be desirable to focus on an intensive research effort in a particular phylum, or on biosynthetic studies dealing with a single species, or on research that concentrates perhaps on a particular class of compounds, or on a given biological activity. The present volume and its successors, therefore, will not adhere to any overall plan. I will attempt to bring together, at convenient intervals, timely and critical reviews that are representative of major current researches and that, hopefully, will also foreshadow future trends. In this way the treatise should remain responsive to the needs of the marine research community. I will be grateful for comments and suggestions that deal with the present or future volumes.

It is indeed a pleasure to acknowledge the cooperation of all workers in the field who have responded so generously and have provided to the individual authors new results prior to publication.

PAUL J. SCHEUER

Preface

The historical development of terrestrial natural product research proceeded, quite naturally, from an investigation of conspicuous organisms and phenomena to a study of those that are not so obvious, that is, from flowering plants to microorganisms. Research into microbial constituents has been richly rewarding, leading to what might be called the penicillin era of natural product research. The relative ease with which microorganisms can be grown in culture, which is in marked contrast to oftentimes difficult and costly procurement of macroorganisms, contributed significantly to the success of this research.

A similar story is beginning to unfold in marine natural products research. Chapter 1 presents the first comprehensive report on the constituents of blue-green algae, a phylum akin to bacteria rather than seaweeds, and one that had not previously been subjected to scrutiny by chemists. The broad spectrum of blue-green algal metabolites and the knowledge that these organisms can be cultured lend themselves to a safe predicition that this will be a rapidly growing segment of marine natural product research. The present account contains much material from the author's laboratory that has not been published elsewhere.

Reviews of guanidine derivatives and of phenols are long overdue, since both structural features occur widely and prominently in marine organisms. The occurrence and chemistry of these metabolites is treated comprehensively and in depth in Chapters 2 and 3. Much material from widely scattered and uncommon sources has been brought together for the first time.

The concluding chapter on chemical marine ecology examines the wide-ranging bases and far-reaching consequences of marine natural product research. It is a reminder to all of us that in the sea (to a much greater extent than on land) all components, living or dead, dissolved, suspended, or afloat, are interdependent. A multidisciplinary approach to marine research is therefore not only meaningful, but necessary.

This volume also contains an Appendix that lists previously known organic compounds that have been newly isolated from marine sources. This list grew out of a discussion at a Gordon Conference, in which the need for such a service was pointed out. I volunteered to contact the research community, from which the modest list in this volume resulted. Your response will determine whether this service is worth continuing.

Once again, my thanks to all researchers who have made results available prior to publication elsewhere.

PAUL J. SCHEUER

CONTENTS OF PREVIOUS VOLUMES

VOLUME I

VOLUME II

VOLUME III

MARINE
NATURAL PRODUCTS

Chemical and Biological Perspectives

Volume IV

Chapter 1

Constituents of Blue-Green Algae

RICHARD E. MOORE

I. INTRODUCTION

The blue-green algae are among the simplest of plants and are so adaptable to extremely different environments that they are ubiquitous. They can be found in polar icecaps and hot springs; in freshwater, brackish water, oceans, or highly saline lakes and seas, and in terrestrial environs ranging from humid areas to deserts. These organisms are not closely related to any other group of algae, but resemble bacteria in their lack of organized nuclei and in their manner of cell division. The affinity of bacteria and blue-green algae, often referred to as cyanobacteria, has been substantiated by modern work (Rippka *et al.*, 1979; Herdman *et al.*, 1979a, b).

1

MARINE NATURAL PRODUCTS

Research on this group has lagged. In their classic treatise on oceanography, Sverdrup *et al.* (1947) remarked that the blue-green algae "are of less general importance in the oceans than are . . . the [other] algal groups." More recently, however, it has been suggested (Fogg *et al.*, 1973) that studies of the blue-green algae might shed light on the origin of life and the evolution of higher plants. Studies have increased appreciably in recent years since certain blue-green algae, like some bacteria, are able to fix nitrogen under aerobic conditions and play important roles in agricultural productivity and numerous ecological situations.

Blue-green algae are of widespread distribution in marine ecosystems but are seldom dominant and generally pass unnoticed by the casual observer. They are fairly abundant in estuarine and intertidal areas where they are subjected to extremes of environmental change. Freshwater species are well known as nuisances and at times have been responsible for animal and even human intoxications. So far, the marine forms have not been as important from an economic and health standpoint. Outside of the work done in the author's laboratory, most of the chemical research reported in the literature has been concerned with the identification of toxic and noxious substances from freshwater species. There are numerous reports of antibacterial, antifungal, antiviral, and other pharmacological activities associated with extracts of freshwater and marine blue-green algae (Baslow, 1977). Two very interesting activities are the healing-promoting effects of extracts of thermophilic algae belonging to the genus *Phormidium* and the carcinogenic nature of extracts of *Nostoc rivulare*. The relationship of the growth-stimulating substances from *Phormidium* to the tumor-inducing factors from *Nostoc* is not clear. The active substances in these cyanophytes and several others have not been isolated and characterized. Although much of the pharmacological evidence is fragmentary, the results obtained to date suggest that future chemical research of the blue-green algae will be exciting and rewarding. The purpose of this chapter is to summarize the chemical work that has been done so far.

A. Phyletic Considerations

Blue-green algae possess prokaryotic cells in which membrane-bound organelles such as nuclei, chloroplasts, and mitochondria are absent. Blue-green algae cells are structurally similar to those of bacteria. All other living organisms have cells that are eukaryotic. The importance of this morphological distinction was presented convincingly by Stanier and van Niel in 1962 in a review entitled "The Concept of a Bacterium" where the bacterial nature of blue-green algae was strongly implied. Actually,

Cohn had recognized a century earlier the close relationship of blue-green algae and bacteria and in 1871–1872 had placed these two prokaryotes in the same phylum Schizophyta, assigning the name Schizophyceae to the blue-green algae and Schizomycetes to the bacteria. His nomenclature, however, never received general acceptance. Blue-green algae shared many features in common with other algal groups, and it was more convenient to treat them as algae. The name Cyanophyta, a term introduced by Smith in 1938, became widely used as it conformed with the rule of naming algal groups according to their characteristic colors. Blue-green algae, however, are not always blue-green in color. As pointed out already, the recent work of Stanier and his group (Rippka et al., 1979; Herdman et al., 1979a,b) leaves little doubt that the blue-green algae are phyletically closely related to bacteria.

B. Biosystematics

Since cyanophtes have been traditionally considered as algae, their classification was developed by phycologists. On the basis of studies made predominantly on field collections, as many as 150 genera and about 1500 species have been described.

Two kinds of blue-green algae are easily discernible: the unicellular coccoid forms and the filamentous forms. The latter group is characterized by the presence of chains of cells known as trichomes. According to classical taxonomy (Desikachary, 1973), the coccoid forms are divided into two orders, the Chroococcales and the Pleurocapsales; and the filamentous forms are divided into two orders, the Nostocales and the Stigonematales (Table 1). In the Chroococcales the plants are unicellular; the cells of a plant, however, can aggregate and form a pseudofilamentous colony, but never one with a trichome organization. Plants belonging to the Pleurocapsales are also unicellular, and the aggregated cells in the thalli can be heterotrichous, filamentous, or nearly filamentous, but not trichomal. In the Nostocales the plants are filamentous and the cells form trichomes; heterocysts, akinetes, and hormogonia can be present; true branching is absent, but false branching can be present. The Stigonematales are heterotrichous filamentous plants that show true branching; possess heterocysts, hormogonia, and sometimes akinetes; and usually exhibit clear pit connections between cells.

Attempts to identify cyanophytes in culture by criteria used in the field often lead to difficulties and ambiguities. To facilitate the generic identification of cultures, Rippka et al. (1979) have revised the criteria and included some that were not previously recognized since they were discovered only as a result of culture investigations. The genera that have been

TABLE 1

Classical Taxonomy of the Cyanophyta[a]

Order	Family	Representative genera
Chroococcales	Chamaesiphonaceae	*Chamaesiphon, Dermocarpa*
	Chroococcaceae	*Chlorogloea, Chroococcus, Coelosphaerium, Gloeocapsa, Gloeothece, Microcystis, Synechococcus, Synechocystis*
Pleurocapsales	Entophysalidaceae	
	Pleurocapsaceae	*Hyella, Myxosarcina, Pleurocapsa, Xenococcus*
	Microchaetaceae	
	Nostocaceae	*Anabaena, Aphanizomenon, Cylindrospermum, Hormothamnion, Nodularia, Nostoc*
Nostocales	Oscillatoriaceae	*Crinalium, Lyngbya, Microcoleus, Oscillatoria, Phormidium, Schizothrix, Spirulina, Symploca*
Stigonematales	Rivulariaceae	*Calothrix, Gloeotrichia, Rivularia*
	Scytonemataceae	*Plectonema, Scytonema, Tolypothrix*
	Capsosiraceae	
	Mastigocladaceae	*Brachytrichia*
	Mastigocladopsidaceae	
	Nostochopsidaceae	*Mastigocladus*
	Stigonemataceae	*Fischerella*

[a] From Desikachary (1973).

studied by Rippka and collaborators have been placed in five sections, each distinguished by a particular pattern of structure and development (Table 2).

II. PIGMENTS

Blue-green algae usually contain chlorophyll a, as do all eukaryotic algae, and biliproteins in their photosynthetic apparatus and carry out photosynthesis with the production of oxygen. Photosynthetic bacteria, however, contain different pigments and do not produce oxygen (Vernon and Seely, 1966).

The major accessory pigment in most blue-green algae is C-phycocyanin, an intensely blue, water-soluble biliprotein composed of two distinct polypeptide chains, α and β subunits (Glazer and Cohen-Bazire, 1971; Bennett and Bogorad, 1971; O'Carra and Killilea, 1971), to which are

TABLE 2

Major Subgroups of Cyanobacteria[a]

Section	Cell arrangement	Type of reproduction	Representative genera
I	Unicellular; single, cells or colonial aggregates held together by additional outer cell wall layers	Binary fission Budding	*Goleobacter, Gloeocapsa, Gloeothece, Synechococcus, Synechocystis* *Chamaesiphon*
II	Unicellular; same as above	Multiple fission only Both binary and multiple fission	*Dermocarpa, Xenococcus Chroococcidiopsis, Dermocarpella, Myxosarcina, Pleurocapsa group*
III	Filamentous; a chain of cells (trichome)	Transcellular or intercellular trichome breakage	*Lyngbya–Phormidium–Plectonema* groups *Oscillatoria, Pseudanabaena, Spirulina*
IV	Filamentous (trichome contains heterocysts)	Random trichome breakage, sometimes by germination of akinetes; division in only one plane As above, but also by formation of hormogonia; division in only one plane	*Anabaena, Cylindrospermum, Nodularia* *Calothrix, Nostoc, Scytonema*
V	Filamentous (trichome contains heterocysts)	Random trichome breakage, formation of hormogonia, sometimes by germination of akinetes; division in more than one plane	*Chlorogloeopsis, Fischerella*

[a] From Rippka *et al.* (1979).

attached three covalently bonded prosthetic groups known as phycocyanobilins, one on the α chain and two on the β chain (Glazer and Fang, 1973). The phycocyanobilins are readily released when a solution of C-phycocyanin in methanol is boiled under reflux. The single pigment that is produced has structure **1** as determined from spectroscopic (Crespi *et al.*, 1967; Cole *et al.*, 1968) and degradative studies (Rüdiger and O'Carra, 1969) and total synthesis (Gossauer and Hinze, 1978). The absolute

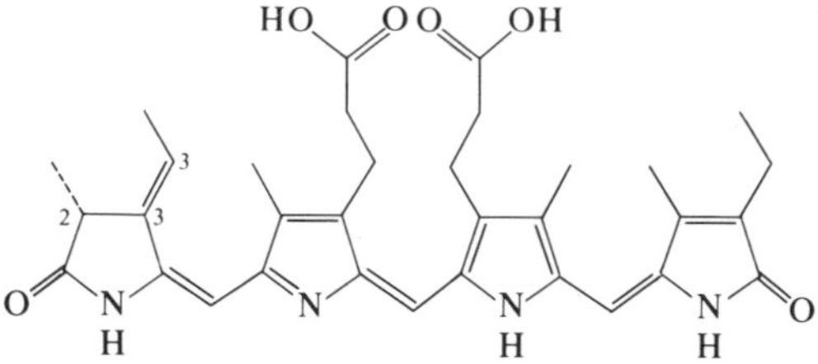

1 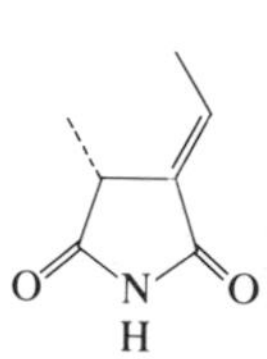   2

stereochemistry at C-2 appears to be *R* since chromate oxidation of C-phycocyanin leads to the levorotatory succinimide **2** (Brockman and Knobloch, 1973). Structure studies of highly purified chromopeptides from degradation of the C-phycocyanins of *Mastigocladus laminosus* (Byfield and Zuber, 1972) and *Synechococcus* sp. 6301 (Williams and Glazer, 1978) have unambiguously established that the three phyco-cyanobilins are attached to cysteine residues in the two polypeptide chains. The smallest fragment, a blue heptapeptide **3** obtained from deg-radation of *Synechococcus* sp. 6301 C-phycocyanin with cyanogen bromide, has been examined by high-frequency nmr spectroscopy and shown to have a thioether functionality connecting the chromophore and peptide (Lagarias *et al.*, 1979). The absolute stereochemistry at C-2, C-3, and C-3' in **3** is probably *RRR*. Since a comparison of the visible spectra of C-phycocyanin and heptapeptide **3** (λ_{max} 660 and 655 nm in 30% HOAc,

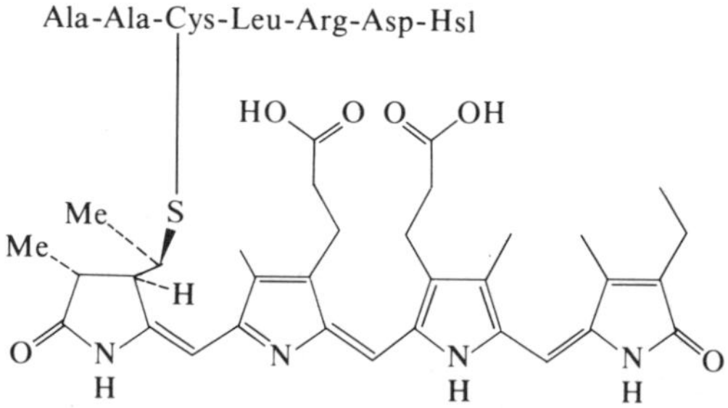

Ala-Ala-Cys-Leu-Arg-Asp-Hsl

3

Hsl = homoserine lactone

respectively) shows insignificant differences, the three phycocyanobilins in C-phycocyanin have the same chromophores as **3**.

Allophycocyanin is a minor biliprotein in most blue-green algae. It consists of a single polypeptide chain with one phycocyanobilin moiety covalently attached to it (Glazer and Fang, 1973). Compound **1** is eliminated when the protein is treated with boiling methanol.

C-Phycoerythrin is the major biliprotein in most red algae, and this purple pigment is sometimes present in blue-green algae. In many species belonging to the family Oscillatoriaceae it is the major biliprotein. Like C-phycocyanin, the protein is composed of two polypeptide chains with three covalently bonded phycoerythrobilin pigments, one on the α chain and two on the β chain (Bennett and Bogorad, 1971). When C-phyco-erythrin is treated with boiling methanol, the phycoerythrobilin **4** is eliminated. The gross structure of **4** was elucidated by spectroscopic (Chapman et al., 1967; Crespi and Katz, 1969) and degradation studies (Rüdiger et al., 1967; Rüdiger and O'Carra, 1969). The E geometry of the ethylidene group attached to C-3 and the R configuration at both C-2 and C-16 in **4** was rigorously established by total synthesis (Gossauer and

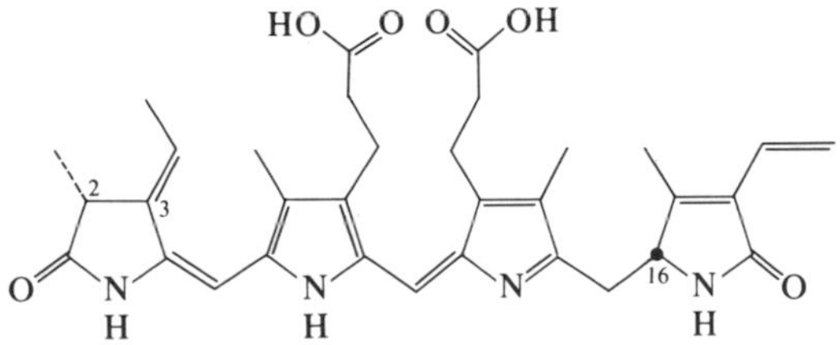

4

Weller, 1978). The chromophoric, prosthetic groups appear to be attached to cysteine residues in the polypeptide chains as shown in **5** (Köst-Reyes et al., 1975).

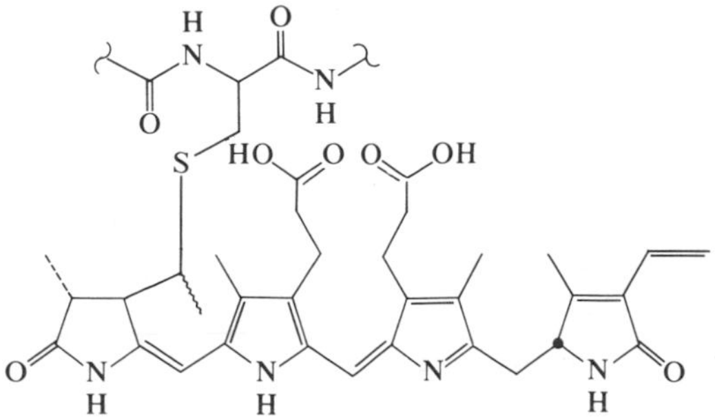

5

Blue-green algae also contain carotenoid pigments. The carotenoid composition of the unicellular blue-green algae has not been studied in detail; much data, however, have been obtained on the carotenoids of the filamentous blue-green algae. β-Carotene is always a major component. The most characteristic carotenoids of blue-green algae are echinenone 4-keto-β-carotene), the monocyclic rhamnoside myxoxanthophyll (6), and the acyclic dirhamnoside oscillaxanthin. These pigments are completely absent in bacteria and, with the exception of echinenone, all other algae. Glycosidic carotenoids, however, are sometimes present in bacteria but have not been found in eukaryotic algae. A noteworthy feature is the total absence of epoxidic carotenoids in blue-green algae. This latter feature is shared with bacteria. For a more detailed presentation of the

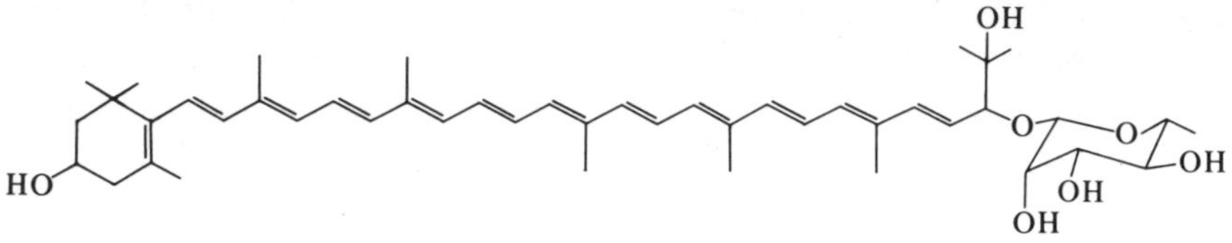

6

structural types of carotenoids found in blue-green algae, the reader is referred to the review by Liaaen-Jensen (1978) in this series.

III. TOXINS

A. From Freshwater Cyanophytes

Blue-green algal blooms are well known as nuisances in freshwater systems and occasionally are responsible for animal and even human intoxications. [For reviews see Moore (1977) and Gorham and Carmichael (1979).] The majority of poisonings have occurred in the lakes, ponds, and reservoirs of the great plains areas of North America and certain strains of *Anabaena flos-aquae*, *Aphanizomenon flos-aquae*, and *Microcystis aeruginosa* have been responsible for most of the outbreaks.

1. Anatoxins from Anabaena flos-aquae

The main toxin associated with *Anabaena flos-aquae* is a water-soluble alkaloid, anatoxin-a (7), which has been shown to be a potent, postsynaptic, depolarizing neuromuscular blocking agent (Carmichael *et al.*, 1975, 1979). The structure and absolute stereochemistry were solved by X-ray

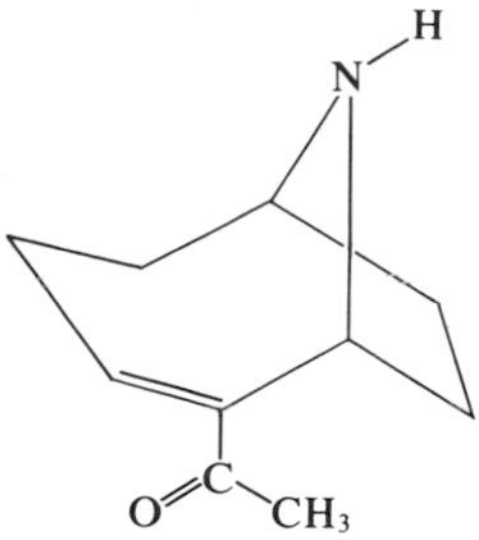

7

crystallography (Huber, 1972) and spectroscopy (Devlin *et al.*, 1977), and verified by a partial synthesis from cocaine (Campbell *et al.*, 1977) as follows: Treatment with hot hydrochloric acid converted cocaine (8) to anhydroecgonine (9). Reaction of the lithium salt of 9 with methyllithium gave the methyl ketone 10. When this ketone was treated with sodium dimethyloxosulfonium methylide in dimethyl sulfoxide, the *endo* cyclopropane 11 was obtained as the major product. Photolysis of 11 led cleanly to N-methylanatoxin-a (12), and anatoxin-a (7) was produced when 12 was reacted with diethyl azodicarboxylate. The synthesis is outlined in Scheme 1.

Anatoxin-a has also been synthesized utilizing an intramolecular cyclization between an iminium salt and a nucleophilic carbon to construct 12 (Bates and Rapoport, 1979) (Scheme 2). Friedel–Crafts acylation of 1-methylpyrrole (13) with the acid chloride of monomethyl glutarate (14) gave predominantly 15. Wolff–Kishner reduction of 15 to 16 followed by treatment of the lithium salt of 16 with methyllithium led to ketone 17. Acylation of 17 with trichloroacetyl chloride gave 18 which reacted with methoxide to produce methyl ester 19. Catalytic hydrogenation of 19 using rhodium/alumina in acidic methanol yielded the pyrrolidine 20. The methyl ketone group was regenerated by Jones oxidation of 20 to 21. The ester 21 was hydrolyzed with aqueous hydrochloric acid and the amino acid 22 was decarbonylated with $POCl_3$ to give the iminium chloride 23. Cyclization of 23 to N-methyldihydroanatoxin-a (24), which had been prepared previously by ring expansion of cocaine (Campbell *et al.*, 1977), was accomplished in acidic methanol. Acetylation converted 24 to a pair of enol acetates (25 and 26) which after bromination and mild hydrolysis afforded the bromoketone 27. Elimination of hydrogen bromide from 27 to give 12 was carried out with lithium bromide and lithium carbonate in dimethylformamide.

Several new toxic clones of *Anabaena flos-aquae* have been grown in axenic culture. The pharmacological properties of the toxic extracts as-

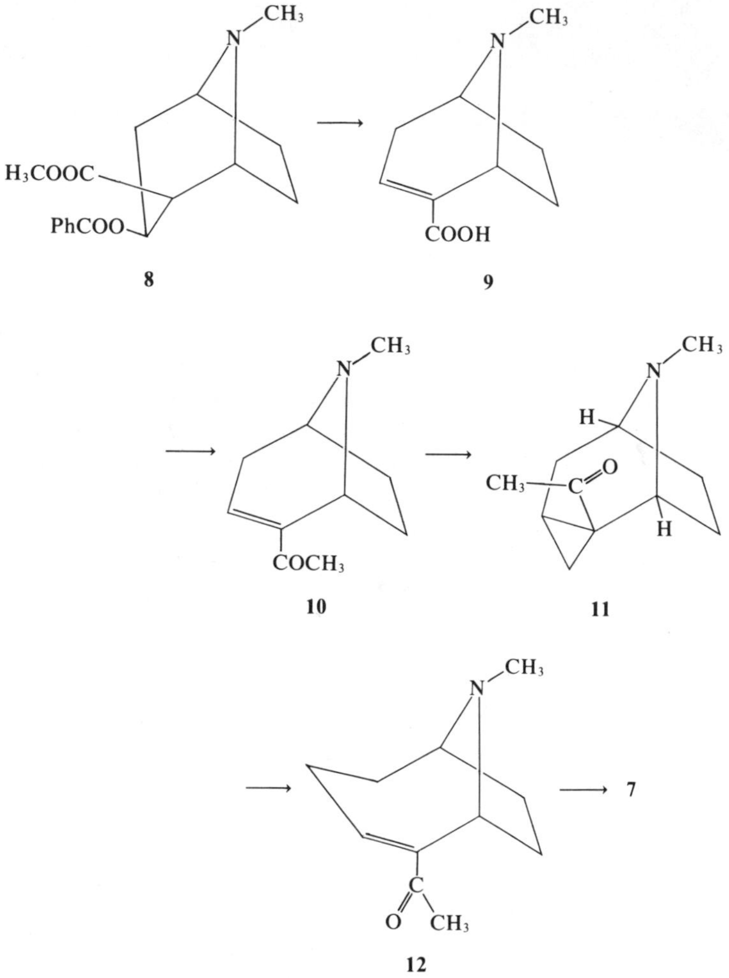

Scheme 1. Synthesis of anatoxin-a (**7**) from cocaine (**8**).

sociated with these clones are different from those of anatoxin-a. Carmichael and Gorham (1978) conclude that there are at least four types of anatoxins (a, b, c, and d) associated with *Anabaena flos-aquae*.

2. Saxitoxins from Alphanizomenon flos-aquae

The toxicity of the blue-green alga *Aphanizomenon flos-aquae* appears to be due to saxitoxin and related compounds of unknown structure (Shimizu, 1978). Saxitoxin (**28**) is generally the major toxin associated with poisonous shellfish such as the Alaskan butter clam, *Saxidomus*

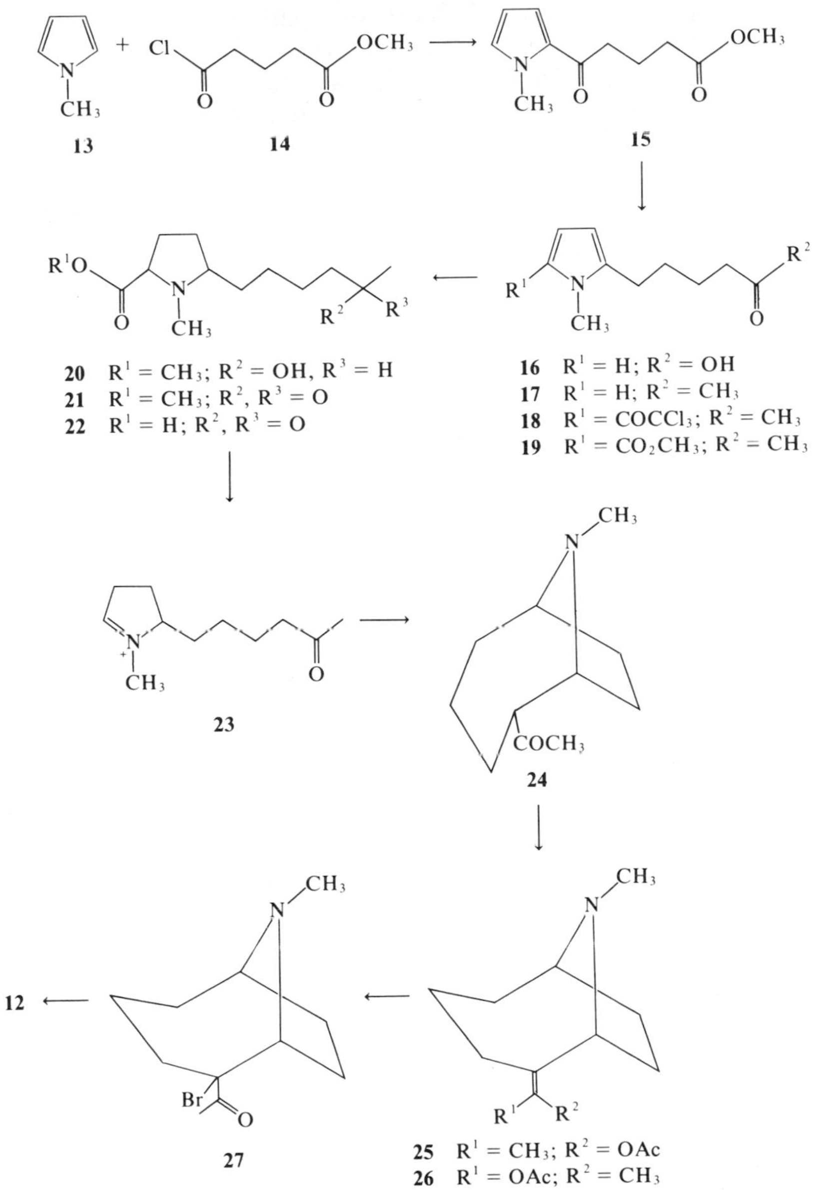

13   **14**   **15**

20 $R^1 = CH_3$; $R^2 = OH$, $R^3 = H$
21 $R^1 = CH_3$; $R^2, R^3 = O$
22 $R^1 = H$; $R^2, R^3 = O$

16 $R^1 = H$; $R^2 = OH$
17 $R^1 = H$; $R^2 = CH_3$
18 $R^1 = COCCl_3$; $R^2 = CH_3$
19 $R^1 = CO_2CH_3$; $R^2 = CH_3$

23   **24**

12   **27**

25 $R^1 = CH_3$; $R^2 = OAc$
26 $R^1 = OAc$; $R^2 = CH_3$

Scheme 2. Synthesis of *N*-methylanatoxin-a via intramolecular cyclization of an iminium salt.

giganteus, and the soft-shell clam, *Mya arenaria* (Shimizu *et al.,* 1978). Saxitoxin accumulates in the siphons of *S. giganteus,* presumably as a result of the ingestion of the toxic dinoflagellate *Gonyaulax catenella;* there is no proof of this however. The origin of the saxitoxin in *M. arenaria* is apparently from the ingestion of *Gonyaulax tamarensis.* Interestingly, neosaxitoxin (**29**) is the major toxin in *G. tamarensis* and saxitoxin is a minor component. Only traces of neosaxitoxin are present in *S. giganteus* and *M. arenaria.* Preliminary biochemical studies indicate that neosaxitoxin, with its easily reducible *N*-hydroxy group, is converted to saxitoxin in the clams. 11α-Hydroxy- (**30,** gonyautoxin II) and 11β-hydroxysaxitoxin (**31,** gonyautoxin III) are also major toxins in *G. tamarensis* (Shimizu *et al.,* 1976); these latter two toxins, however, probably exist as the C-11 sulfate esters **32** and **33** in the dinoflagellate (Boyer *et al.,* 1978).

28	$R^1, R^2, R^3 = H$	**31**	$R^1, R^3 = H; R^2 = OH$
29	$R^1 = OH; R^2, R^3 = H$	**32**	$R^1, R^2 = H; R^3 = OSO_3^-$
30	$R^1, R^2 = H; R^3 = OH$	**33**	$R^1, R^3 = H; R^2 = OSO_3^-$

The structure and absolute configuration of saxitoxin have been rigorously established by X-ray crystallography (Schantz *et al.,* 1975; Bordner *et al.,* 1975).

A stereospecific total synthesis of *d,l*-saxitoxin has been reported (Tanino *et al.,* 1977) and is outlined in Scheme 3. Methyl 2-oxo-4-phthalimidobutyrate was converted to ketal **34** with 1,3-propanediol in the presence of *p*-toluenesulfonic acid. Reaction with hydrazine liberated an intermediate primary amine which cyclized to lactam **35**. Compound **35** was then reacted with phosphorus pentasulfide to yield thiolactam **36**. Synthesis of the vinylogous carbamate **37** was accomplished by condensing **36** with methyl 2-bromoacetoacetate in the presence of NaHCO₃ and degrading the resulting keto ester with methanolic KOH. Carbamate **37** was condensed with benzyloxyacetaldehyde and silicon tetraisothiocyanate to give the thiourea ester **38**. The hydrazide **39** was formed by treating **38** with hydrazine and transformed into an azide **40** by the action of NOCl. Pyrolysis of **40** produced an isocyanate **41**, which added am-

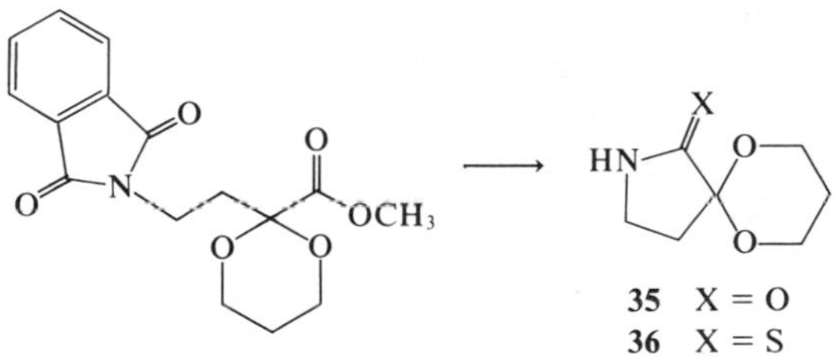

35 X = O
36 X = S

34

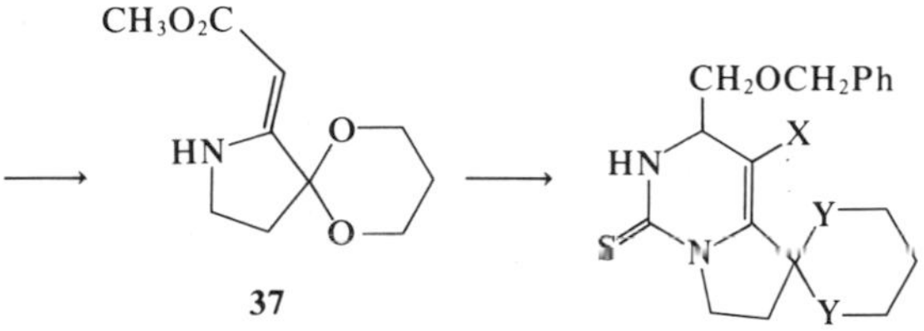

37

CH_2OCH_2Ph

38 X = CO_2CH_3; Y = O
39 X = $CONHNH_2$; Y = O
40 X = CON_3; Y = O
41 X = NCO; Y = O
42 X = $NHCONH_2$; Y = O
43 X = $NHCONH_2$; Y = S

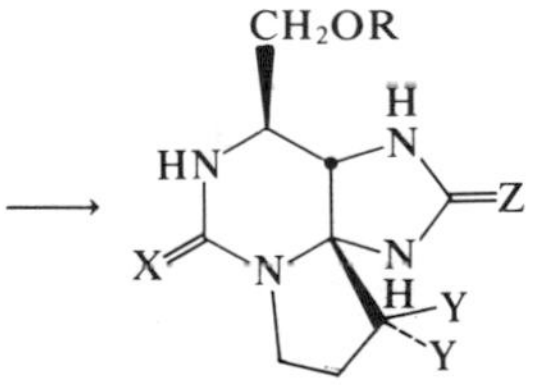

CH_2OR

44 X = S; Y = $S(CH_2)_3S$; Z = O; R = CH_2Ph
45 X, Z = NH; Y = $S(CH_2)_3S$; R = CH_2Ph
46 X, Z = NH; Y = $S(CH_2)_3S$; R = H
47 X, Z = NH; Y = OH, OH; R = H
28 X, Z = NH; Y = OH, OH; R = $CONH_2$

Scheme 3. Synthesis of *d,l*-saxitoxin.

monia to give the thiourea urea **42**. Compound **42** was converted to the thioketal thiourea **43** and cyclized to **44** in the presence of a 9:1 mixture of acetic acid and trifluoroacetic acid. The tricyclic thiourea **44** was transformed into the diguanidine **45** by reacting **44** with boron trifluoride etherate in the presence of sodium bicarbonate and then heating the intermediate diurea with ammonium propionate. The hydrochloride salt of **45** was treated with boron trichloride to give decarbamoylsaxitoxin thioketal **46** which was isolated as the hexaacetate. Treatment of this hexacetate with N-bromosuccinimide in wet acetonitrile followed by methanolysis gave decarbamoylsaxitoxin (**47**) isolated as the dihydrochloride. Reaction of **47** with chlorosulfonyl isocyanate in formic acid followed by hot water workup led to d,l-saxitoxin.

3. Toxic Peptides from Microcystis aeruginosa

The toxins associated with *Microcystis aeruginosa* appear to be polypeptides (Bishop *et al.*, 1959; RamaMurthy and Capindale, 1970; Kirpenko *et al.*, 1975). Runnegar and Falconer (1975) have isolated a toxic peptide from *M. aeruginosa* by alkaline extraction of lyophilized alga. The toxin has no free amino group and on hydrolysis equimolar amounts of L-methionine, L-tyrosine, D-alanine, D-glutamic acid, *erythro-β*-methylaspartic acid, and methylamine are obtained. The LD_{50} of the purified toxin is 0.056 mg/kg in mice.

B. From Marine Cyanophytes

Although marine cyanophytes have not presented serious health and economic problems to date, the potential danger is there, especially as man increases his utilization of the ocean as a food source. A recent striking example occurred in 1975, 1976, and 1977 at one of the world's largest intensive-culture shrimp research projects located at Puerto Peñasco, Mexico. Blooms of a marine species of a blue-green alga, first identified as *Spirulina subsalsa* (Lightner, 1978) and later as a *Lyngbya* sp. (D. V. Lightner, unpublished work, 1979), were found to be related to a particular disease syndrome in raceway-reared blue shrimp (*Penaeus stylirostris*). This highly fatal disease, characterized by necrosis of the lining epithelium of the midgut, dorsal cecum, and hindgut gland, and a consequent hemocytic enteritis in the shrimp, was probably caused by a highly inflammatory agent in the *Lyngbya*.

The first marine cyanophyte to be examined in the author's laboratory was a toxic, deep-water variety of *Lyngbya majuscula** from Enewetak Atoll in the Marshall Islands. A strain of this alga had previously been shown to be the causative agent of "swimmers' itch," a severe contact dermatitis afflicting swimmers in Hawaii during the summer months (Banner, 1959). The toxin in the *L. majuscula* from Enewetak, as well as the inflammatory agent in a shallow-water, dermatitis-producing strain from Laie, Oahu, was shown to be debromoaplysiatoxin (Mynderse *et*

*This cyanophyte was erroneously identified as *Lyngbya gracilis* by Mynderse *et al.* (1977). *L. majuscula* is sometimes called *Microcoleus lyngbyaceus* (Drouet, 1968), which the author considers to be a less desirable binomial. Drouet's taxonomic revisions are oversimplified and therefore unacceptable to many biologists (Desikachary, 1973).

al., 1977), one of the poisonous substances that Kato and Scheuer (1975) had isolated from the digestive gland of the sea hare *Stylocheilus longicauda*. This was the first time that a toxin had been identified from a marine blue-green alga. Debromoaplysiatoxin was a highly inflammatory agent which caused an erythematous pustular folliculitis in humans (Solomon and Stoughton, 1978). The vesicatory substance in a shallow-water variety of *L. majuscula* found at Kahala Beach, Oahu, however, was an indole alkaloid, lyngbyatoxin A (Cardellina *et al.*, 1979), which was structurally and pharmacologically closely related to teleocidin B, an inflammatory agent found in the mycelia of several strains of *Streptomyces* (Sakabe *et al.*, 1966). Teleocidin B had been reported to cause intense irritation on rabbit skin (Takashima *et al.*, 1962) and severe irritation and eruptive vesications on human skin (Nakata *et al.*, 1966). Lyngbyatoxin A produced similar effects.

The chloroform extracts of the deep-water varieties of *L. majuscula* from Reefer 8 pinnacle and South Elmer pinnacle, Enewetak, and the shallow-water varieties of *L. majuscula* from Laie and Kahala Beach, Oahu, consistently display activity against P-388 lymphocytic mouse leukemia. The active principle in the first three varieties has been shown to be debromoaplysiatoxin (Mynderse *et al.*, 1977) whereas the antileukemic agent in the fourth variety is lyngbyatoxin A (Cardellina *et al.*, 1979c). No anticancer activity is associated with the aqueous extracts of these four varieties of *L. majuscula*. The aqueous extracts of two other specimens of *Lyngbya* (a shallow-water *Lyngbya* sp. from the City of Refuge, Hawaii and a variety of *L. majuscula* from Palau) show good activity against P-388 leukemia and also completely inhibit Ehrlich ascites tumor in mice (Table 3). Lipophilic and hydrophilic extracts of several other blue-green algae from the Pacific basin also show good activity against P-388 leukemia and/or Ehrlich ascites tumor. Generally much less chronic toxicity is associated with the anticancer compounds in the hydrophilic extracts. The anticancer agents in the lipophilic extracts of a mixture of *Oscillatoria nigroviridis* and *Schizothrix calcicola* from Enewetak and a *Tolypothrix conglutinata* var. *clorata* from Fanning Island have been isolated and characterized. Very little is known at present, however, about the anticancer compounds in the remaining cyanophytes listed in Table 3.

Debromoaplysiatoxin and lyngbyatoxin A are not effective anticancer drugs. The highest activity against P-388 leukemia is obtained only at or near the chronic toxicity level. Debromoaplysiatoxin shows no activity in the B16 melanoma and Lewis lung tumor test systems.

Interestingly, both debromoaplysiatoxin and lyngbyatoxin A are potent tumor promoters (H. Fujiki and T. Sugimura, unpublished work, 1980). Topical application of either compound on mouse skin results in a rapid,

TABLE 3
Anticancer Activities of Blue-Green Algal Extracts[a]

| Binomial of alga | Location | Extracting solvent and purity of fraction[b] | P-388 Lymphocytic leukemia[c] | | | Ehrlich ascites[d] | | |
| | | | Dose (mg/kg) | T/C Activity (%) | No. of mice[e] | Dose (mg/kg) | Survivors at 30 days | |
							Alive (%)	Nonascitic (%)
Crinalium sp.	Fanning Is.	EtOAc, I	17.5	178	6(1)	14.0	80	80
		H₂O, III	0.5	210	4			
Lyngbya convervoides	Palau	30% EtOH, I				20	100	100
Lyngbya majuscula	Reefer 8 pinnacle, Enewetak	CHCl₃, I	1.5	167	5			
Lyngbya majuscula	South Elmer, pinnacle, Enewetak	CHCl₃, I	0.65	137	5			
Lyngbya majuscula	Kahala Beach, Oahu	CHCl₃, I	0.5	144	5	0.4	0	0
Lyngbya majuscula	Palau	H₂O, I	25.0	166	5(1)	8	60	40
		H₂O, II	10.0	164	5	<0.4	100	100
		H₂O, III	<0.5	166	5			

transient stimulation of epidermal ornithine decarboxylase (ODC) activity, an activity which is elevated in fast-growing neoplasms such as L1210 leukemic cells (Russell and Levy, 1971). Both toxins also induce cell adhesion of human promyelocytic leukemia cells (HL-60) and inhibit terminal differentiation of Friend erythroleukemia cells induced by dimethyl sulfoxide. The doses of lyngbyatoxin A required for these effects are similar to, or a little lower than, those of the poten tumor promoter 12-O-tetradecanoylphorbol-13-acetate (TPA) from Croton oil and comparable to those of teleocidin B and dihydroteleocidin B (Fujiki *et al.*, 1979). The dose of debromoaplysiatoxin necessary for ODC activity is a little lower than that of TPA; the doses required to induce cell adhesion of HL-60 cells and to inhibit terminal differentiation of Friend erythroleukemia cells, however, is an order of magnitude higher.

The experimental induction of mouse skin tumors proceeds in at least two stages: initiation and promotion. Following the application of a single subcarcinogenic dose of an initiator such as 7,12-dimethylbenz[a] anthracene, tumors can be obtained with repetitive topical applications of TPA (Baird and Boutwell, 1971) or dihydroteleocidin B (H. Fujiki and T. Sugimura, unpublished results, 1980). Similar *in vivo* experiments on the effect of painting mice with a limited amount of 7,12-dimethylbenz[a] anthracene followed by debromoaplysiatoxin or lyngbyatoxin A are in progress. These compounds may play a role in the development of human cancer. *Lyngbya majuscula,* for example, is sometimes found as an epiphyte on edible seaweeds such as *Acanthophera spicifera* (Rhodophyta).

1. Aplysiatoxins

Watson (1973) found that the digestive glands of four species of sea hares from Hawaii, *Dolabella auricularia, Aplysia pulmonica, Stylocheilus longicauda,* and *Dolabrifera dolabrifera,* contained two distinct toxins, an ether-soluble toxin and a water-soluble toxin, with clearly different physiological properties (Watson and Rayner, 1973). Kato and Scheuer (1975) isolated the ether-soluble toxin of *S. longicauda* and showed that it was predominantly a mixture of two toxic substances

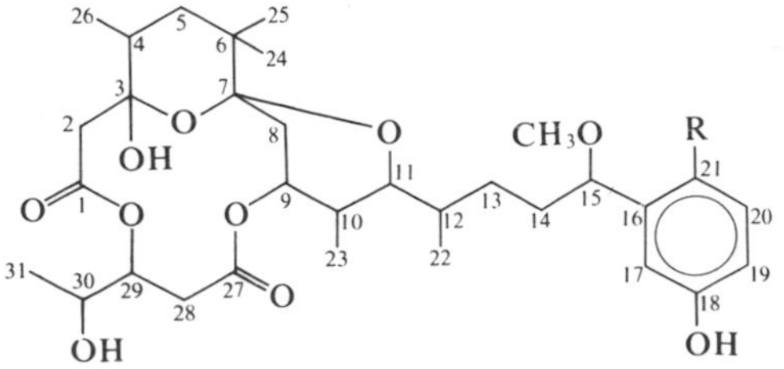

48   R = Br
49   R = H

		[b]					[e]	
Lyngbya sp.	City of Refuge, Is. of Hawaii, Hawaii	MeOH, I	33	136	5	27.2	100	100
		MeOH, I				1.6	100	80
		MeOH, II						
Oscillatoria sp.	Palau	30% EtOH, I	150	182	5	4.0	100	100
		30% EtOH, III	5	205	5	0.4	100	100
Oscillatoria nigroviridis Schizothrix calcicola[f]	Enewetak	CHCl₃, 1	0.5	164	5	10	100	100
Phormidium sp.	Molokai, Hawaii	MeOH, I	12.5	210				
Phormidium crosbyanum	Johnston Is.	CH₂Cl₂, I	7.5	134	5(3)	6	0	0
Schizothrix calcicola	Fanning Is.	EtOAc, I	25	255	5(3)	2.5	100	100
		H₂O, II	2.5	209	5(1)	3.6	100	80
Tolypothrix conglutinata var. *clorata*	Fanning Is.	EtOAc, I	13.5	233	5(3)	11	0	0

[a] From Kashiwagi et al. (1980).

[b] I, Crude; II, fraction chromatographed through Sephadex G-25; III, fraction chromatographed through CM Sephadex C-25.

[c] Procedure described by Dunn et al. (1975).

[d] Procedure described by Tabrah et al. (1972).

[e] Number in parenthesis is the number of mice that died within 7 days after tumor cell inoculation; the death times of these mice were not used in the calculation of the percentage of activity.

[f] This algal pair has also been identified as *Oscillatoria* sp. and *Crinalium* sp.

which they named aplysiatoxin (48) and debromoaplysiatoxin (49). (Interestingly, the digestive tract of sea hares is unaffected by these highly inflammatory agents.) The absence of these toxins in *S. longicauda* raised in an aquarium suggested that the toxins had a dietary origin. The Scheuer group (Rose *et al.*, 1978) had observed *S. longicauda* feeding on the red alga *Acanthophera spicifera* and, to a lesser extent, on *L. majuscula*. They were unable to find aplysiatoxins in the red alga and did not investigate *L. majuscula* as it was not very abundant at their principal collection site. Mynderse *et al.* (1977), however, were able to isolate debromoaplysiatoxin, but not aplysiatoxin, from *L. majuscula;* during the collection, several *S. longicauda* were found grazing on the blue-green alga, providing strong evidence that the sea hare toxins are obtained from diet.

The gross structures of aplysiatoxin and debromoaplysiatoxin were elucidated from chemical and spectral data (Kato and Scheuer, 1975). A review of the structure determination has already been presented in this series (Moore, 1978). Kato and Scheuer (1976) concluded from nmr evidence and chemical data that the relative stereochemistry of aplysiatoxin (48a) and debromoaplysiatoxin (49a) is $3R^*, 4S^*, 7S^*, 9R^*, 10R^*, 11S^*$; stereochemical assignments, however, could not be made at C-12, C-15, C-29, and C-30.

48a   R – Br
49a   R – H

Mynderse *et al.* (1977) were able to crystallize debromoaplysiatoxin from aqueous methanol. The crystals, which melted at 105.5°–107.0°C, had orthorhombic symmetry, but the unit cell unfortunately consisted of two molecules, and at this writing, a solution of the structure has not been achieved (J. Clardy, personal communication, 1979). Anhydrodebromoaplysiatoxin (50), the major product resulting from dehydration of debromoaplysiatoxin in chloroform solution, could also be crystallized (mp 116.0° −117.5°C), but again the crystals were unsuitable for X-ray analysis.

The circular dichroism (CD) spectrum of debromoaplysiatoxin in ethanol, which has an optical rotation $[\alpha]_D$ +60.6° in ethanol (Mynderse and Moore, 1978a), shows a positive Cotton effect with maxima at 286 and 269 nm having molecular ellipticities of 1031 and 902, respectively, for the ^1L_b transition of the benzene chromophore (R. W. Woodard, J. C. Craig,

R. E. Moore, and J. S. Mynderse, unpublished work, 1979). Similar CD spectra were given by R-($-$)-noradrenaline hydrochloride (**51**), the cactus alkaloid R-($-$)-calipamine hydrochloride (**52**), S-($-$)-1-phenylethanol (**53**) S-($-$)-1-methoxy-1-phenylethane (**54**), and R-($-$)-1-phenylethane-1,2-diol (**55**). The absolute configuration of C-15 in debromoaplysiatoxin (**49b**) is therefore S.

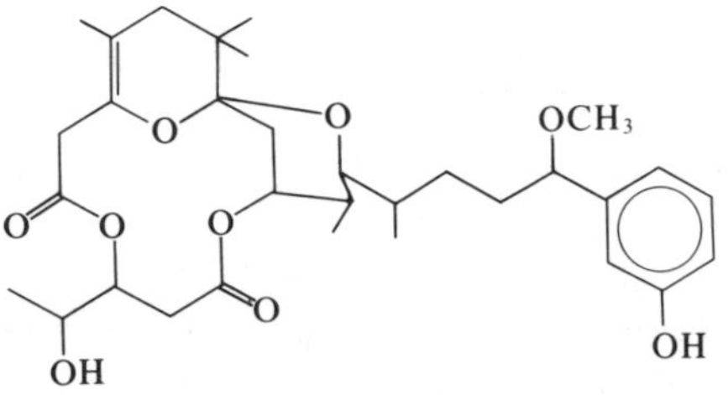

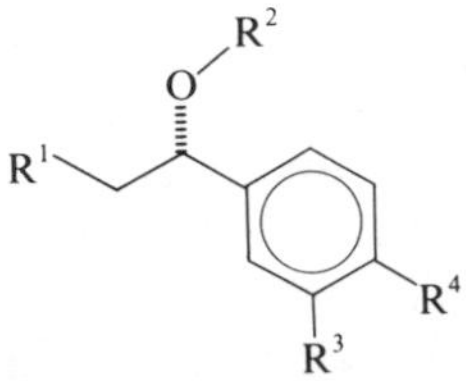

	R^1	R^2	R^3	R^4
49b	$C_{23}H_{37}O_8$	CH_3	OH	H
51	$N^+H_3CH_2$	H	OH	OH
52	$Cl^-CH_3N^+H_2CH_2$	CH_3	OCH_3	OCH_3
53	H	H	H	H
54	H	CH_3	H	H
55	OH	H	H	H

Debromoaplysiatoxin is also a major toxin in a mixture of predominantly two cyanophytes, tentatively identified as *Oscillatoria nigroviridis* and *Schizothrix calcicola,* from the seaward side of Enewetak (Mynderse *et al.*, 1977). Aplysiatoxin (**48**) was not identified in this algal mixture, but a small amount of 19-bromoaplysiatoxin (**56**) was shown to be present (Mynderse and Moore, 1978a).

While investigating possible sources of the poison in ciguateric fish of the Gilbert Islands, Banner (1967) found that two lipid-soluble toxins and a water-soluble toxin were present in *Schizothrix calcicola* from Marakei Atoll, but none of the toxins were characterized and both lipophilic toxins proved to be nonciguateric. The two lipophilic toxins, however, may be related to the aplysiatoxins or oscillatoxins. Interestingly, both an *Oscillatoria* sp. (Schwimmer and Schwimmer, 1964) and *Shizothrix calcicola* (Libby and Erb, 1976) have been implicated as causative agents of human gastrointestinal disorders obtained from drinking water that had become contaminated with these cyanophtes. Keleti *et al.* (1979) suggest that the toxins associated with these freshwater strains are lipopolysaccharides similar to the endotoxins of gram negative bacteria.

2. Oscillatoxins

Mynderse and Moore (1978a) have isolated and identified a second major toxic component, oscillatoxin A (**57**), from the *Oscillatoria* *nigroviridis–Schizothrix calcicola* mixture from Enewetak. The structure of **57** was deduced by comparing the ¹H nmr (Table 4) and ¹³C nmr data of **49** and **57** which clearly established that C-31 was missing in **57** and that **57** was simply 31-nordebromoaplysiatoxin. The 31-nor compound had a comparable optical rotation, $[\alpha]_D$ +67±10° in ethanol, and exhibited the same CD curve as **49,** suggesting that the two toxins have the same absolute stereochemistry. The toxicities and anticancer activities of **57** and **49** were also found to be identical.

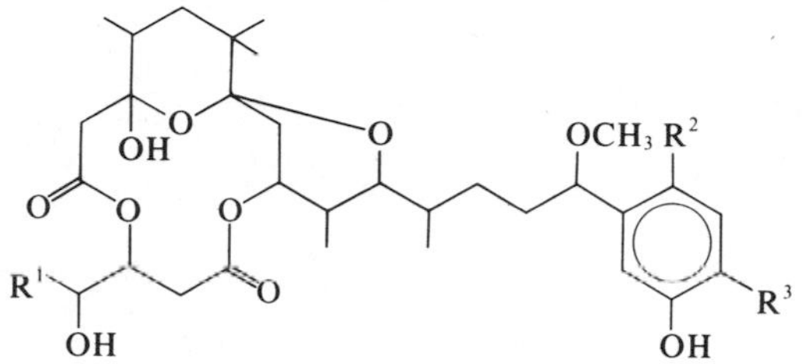

	R¹	R²	R³
56	CH₃	Br	Br
57	H	H	H
58	H	Br	H
59	H	Br	Br

Small amounts of 21-bromo- and 19,21-dibromooscillatoxin A (**58** and **59**) were also found in this algal mixture. In addition minor amounts of **50** and anhydrooscillatoxin A have also been detected and small amounts of four new related compounds, viz., oscillatoxin B (OT-B), 31-noroscilla-toxin B (31-norOT-B), oscillatoxin C (OT-C), and oscillatoxin D (OT-D) have been isolated (Mynderse and Moore, unpublished work, 1978). The structures of the latter four compounds are uncertain at this writing.

The gross structures of OT-B and 31-norOT-B may be **60** and **61,**

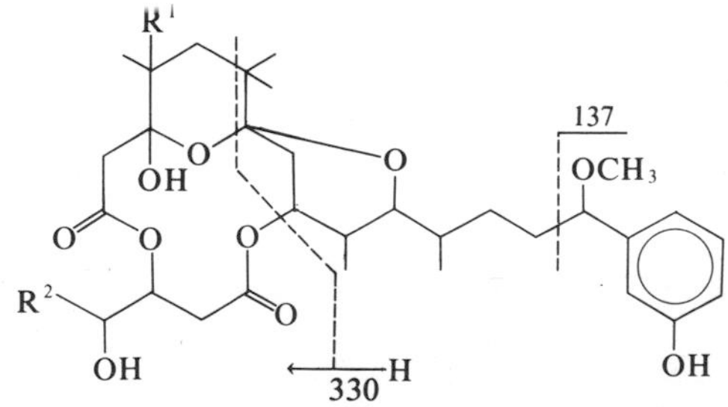

	R¹	R²
49	H	CH₃
57	H	H
60	OH	CH₃
61	OH	H

TABLE 4

Proton nmr Data for Debromoaplysiatoxin (49), Oscillatoxin A (57), Anhydrodebromoaplysiatoxin (50), and Debromoaplysiatoxin Diacetate in Acetone-d_6

Assignment[a]	δ in ppm[b] (Multiplicity; J in Hz)			
	49	57	18,30-Diacetate of 49	50
2	2.77 (d; 13) 2.53 (d; 13)	2.74 (d; 13) 2.53 (d; 13)	2.84 (d; 12.5) 2.48 (d; 12.5)	3.33 (d; 14) 3.07 (br d; 14)
OH on 3	4.30 (d; 2)	4.35 (d; 2)	4.24 (d; 2)	
4	1.86 (m)	1.84 (m)	1.86 (m)	
5	1.63 (t; 13)	1.62 (t; 13)	1.63 (t; 13)	2.20 (br d; 16) 1.36 (br d; 16)
8	1.06 (dd; 13, 4) 2.70 (dd; 14.5, 3) 1.71 (dd; 14.5, 3.5)	1.06 (dd; 13, 4) 2.72 (dd; 14.5, 3) 1.72 (dd; 14.5, 3.5)	1.08 (dd; 13, 4) 2.69 (dd; 15, 3) 1.72 (dd; 15, 3.5)	2.23 (dd; 14.5, 2.5) 1.74 (dd; 14.5, 3.5)
9	5.24 (m)	5.21 (m)	5.27 (br q; 3–3.5)	4.86 (br q; 2.5–3.5)
10	1.7 (m)	1.70 (m)	1.7 (m)	1.72 (m)
11	3.94 (dd; 10.5, 2.5)	3.92 (dd; 10.5, 2.5)	3.95 (dd; 10.5, 2)	3.78 (dd; 11, 2)
12	1.53 (m)	1.53 (m)	1.55 (m)	1.52 (m)
13	1.3–1.4 (m)	1.40 (m) 1.32 (m)	1.3–1.4 (m)	1.47 (m)
14	1.98 (tt; 13, 6.5) 1.6 (m)	1.95 (tt; 13, 6.5) 1.60 (m)	2.00 (m) 1.6 (m)	1.74 (m)
15	4.02 (t; 6.5)	3.99 (t; 6.5)	4.11 (t; 6.5)	4.01 (t; 6.5)
17	6.95 (dd; 2, 1)	6.93 (dd; 2, 1)	7.19 (dd; 2, 1)	6.87 (dd; 2, 1)

OH on 18				
OAc on 18	8.21 (s)	8.25 (s)	2.27 (s)	8.18 (s)
19	6.74 (ddd; 7.5, 2, 1)	6.72 (ddd; 7.5, 2, 1)	7.04 (ddd; 7.5, 2, 1)	6.73 (ddd; 7.5, 2, 1)
20	7.16 (t; 7.5)	7.13 (t; 7.5)	7.38 (t; 7.5)	7.17 (t; 7.5)
21	6.86 (dt; 7.5, 1)	6.84 (dt; 7.5, 1)	7.27 (dt; 7.5, 1)	6.83 (dt, 7.5, 1)
22	0.80 (d; 7)	0.80 (d; 7)	0.80 (d; 7)	0.84 (d; 7)
23	0.72 (d; 7)	0.73 (d; 7)	0.73 (d; 7)	0.82 (d; 7)
24, 25	0.85 (s) / 0.82 (s)	0.84 (s) / 0.81 (s)	0.85 (s) / 0.81 (s)	0.96 (s) / 0.83 (s)
26	0.88 (d; 7)	0.36 (d; 7)	0.90 (d; 7)	1.60 (br s)
28	2.93^c / 2.91^c	2.96^c / 2.93^c	3.01 (dd; 18, 11.5) / 2.84 (dd; 18, 2)	2.77 (dd; 18, 2) / 2.74 (dd; 18, 10)
29	5.25 (m)	5.22 (m)	5.41 (ddd; 11, 5 3.5, 2)	5.35 (dt)
30	4.07 (br m)	3.69 (m) / 3.67 (m)	5.15 (qd; 6.5, 4)	3.86 (m)
OH on 30	4.23 (br m)	4.14 (t; 6.5)		4.00 (d; 5.5)
OAc on 30			2.05 (s)	
31	1.14 (d; 6.5)		1.22 (d; 7)	1.11 (d; 6.5)
OCH₃	3.19 (s)	3.17	3.21 (s)	3.18 (s)

^{a} Based on extensive spin–spin decoupling experiments at 360 MHz.

^{b} Relative to peak for residual acetone-c_5 (δ 2.06) as internal standard.

^{c} AB part of ABX spectrum; $J_{gem} = 18$ Hz.

respectively. The electron-impact (EI) mass spectra of OT-B and 31-norOT-B show intense fragment ion peaks at m/e 330, 298 (330 − CH₃OH), and 137. The same intense fragment ion peaks are exhibited in the EI mass spectra of **49** and **57,** suggesting that the moiety represented by carbons 6 to 25 in **49** and **57** is present in OT-B and 31-norOT-B. Analysis of the ¹H nmr spectra of OT-B and 31-norOT-B confirms this structural conclusion. The ¹H nmr data, shown in Table 5, are consistent with structures **60** and **61,** respectively, except for the following discrepancies. First, it is not certain whether the 2H singlet at δ 2.88 in either spectrum represents the C-2 protons or water in the solvent. Second, it is not certain whether the sharp singlet at δ 5.42 in the OT-B spectrum or at δ 5.38 in the 31-norOT-B spectrum is due to an impurity or a nonexchangeable proton in the molecule; it does not disappear on the addition of D₂O. Third, the singlet at δ 10.38 in the OT-B spectrum or at δ 10.42 in the 31-norOT-B spectrum, which disappears on the addition of D₂O, suggests the presence of an acidic or strongly hydrogen-bonded OH group in both compounds. Finally, only two OH signals appear to be present in either spectrum. A compound having a related structure **62,**

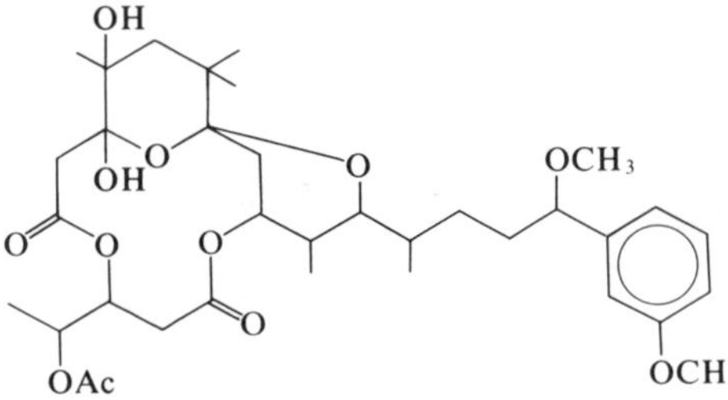

62

obtained by reacting anhydrodebromoaplysiatoxin methyl ether acetate with osmium tetroxide, has been described by Kato and Scheuer (1976).

Structures **63** and **64** for OT-B and 31-norOT-B, respectively, also fit the ¹H nmr data. Ultraviolet spectral data, however, do not support the

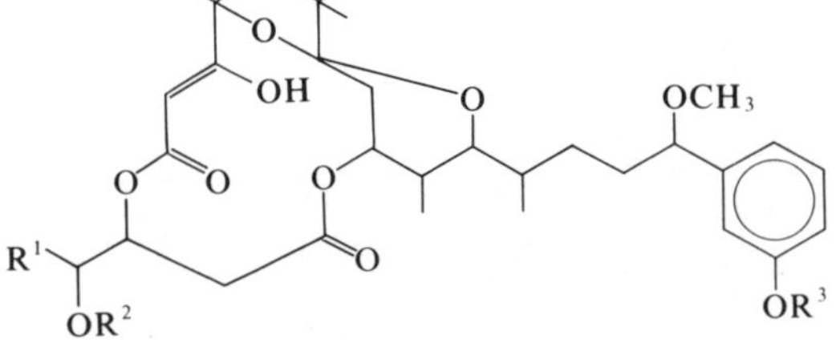

	R¹	R²	R³
63	CH₃	H	H
64	H	H	H
65	CH₃	Ac	CH₃

TABLE 5

Proton nmr Data for Oscillatoxin B and 31-Noroscillatoxin B in Acetone-d_6

δ, ppm[a]		Number of protons	Assignment[b]	Multiplicity; $|J|$ in Hz
OT-B	31-NorOT-B			
10.38	10.42	1	OH	
8.34	8.36	1	OH and 18	
7.16	7.16	1	20	
6.84	6.84	1	17	
6.81	6.81	1	21	
6.73	6.73	1	19	
5.42	5.38			s
5.15		1	29	dt; 10, 4
	5.09	1	29	tt
4.81	4.80	1	9	br quartet; 2.5–3.5
4.11		1	30	qd; 7, 4
4.03	4.03	1	15	dd[c]
	3.71	2	30	d; 4
3.71	3.68	1	11	br d; 10.5
3.16	3.16	3	OCH$_3$	s
2.88	2.88	2	2	s
	2.85	2	28	d; 7
2.79		1	28	dd; 18, 4
2.73		1	28	dd; 18, 10
2.39	2.41	1	8 (eq)	dd; 15.5, 2.5
2.07	2.09	1	5	d; 15
1.7–1.8	1.7–1.8	1	10	m
1.7–1.8	1.7–1.8	2	14	m
1.74	1.76	1	8 (ax)	dd; 15.5, 3.5
1.5–1.6	1.5–1.6	1	12	m
1.58	1.58	1	5	d; 15
1.3–1.4	1.3–1.4	2	13	m
1.38	1.38	3	26	s
1.10		3	31	d; 6.5
1.08	1.08	3	25	s
0.92	0.92	3	24	s
0.86	0.86	3	23	d; 7
0.82	0.82	3	22	d; 7

[a] Relative to peak for residual acetone-d_5 (δ 2.06) as internal standard.
[b] Based on extensive spin–spin decoupling experiments at 220 MHz.
[c] X part of ABX spectrum $J_{AX} + J_{BX} = 13$.

presence of a β-keto ester chromophore. The uv spectra of OT-B and 31-norOT-B in neutral and basic media are essentially identical with those of **49** and **57**. Kato (1973) has proposed that **62** rearranges in dilute acetic acid to a compound having structure **65**. The rearrangement product, however, was not completely characterized; the uv spectrum, for example, was not determined.

The EI mass spectra of OT-B and 31-norOT-B show small ion peaks at *m/e* 590 and 576, respectively, which should be the molecular ions if **63** and **64** are the correct structures or M − H_2O peaks if **60** and **61** are the correct structures. Unfortunately, both the OT-B and 31-norOT-B decomposed before further work could be done to solve the structures.

Oscillatoxin C might be a stereoisomer of **49**. The EI mass spectrum of OT-C is similar to that of **49**, showing intense fragment ion peaks at *m/e* 330, 298, and 137 and a small ion peak at *m/e* 574. The ¹H nmr spectrum of OT-C in acetone-d_6 is similar to that of **49**, the same signal patterns can be seen, but the chemical shifts are quite different. The signals for five of the six methyl groups, for example, resonate at lower field.

Oscillatoxin D appears to have a structure that is related to **66** from a perusal of its ¹H nmr spectrum. The ¹³C nmr spectrum indicates that OT-D

66   **67**

has 32 carbons. Only five methyl carbon signals, however, are observed; C-31 is not a methyl carbon in this compound. Three carbonyl carbon signals are present, of which two are ester type and one is a ketone type. There are no signals for quaternary carbons bearing two oxygens, but there is a signal for a quaternary carbon with one oxygen attached to it. OT-D is unstable; in benzene at room temperature, for example, it is slowly, but cleanly converted to a closely related substance. A detailed study of this rearranged OT-D at 360 MHz in benzene-d_6 suggests that it has partial structure **67** (Table 6).

3. Lyngbyatoxins

Lyngbyatoxin A (**68**) represents the first indole alkaloid from a marine plant. The toxin was isolated in 0.02% yield from a shallow-water variety of *Lyngbya majuscula* found at Kahala Beach, Oahu (Cardellina *et al.*, 1979c). The gross structure was established by ultraviolet, infrared, mass, and detailed high-frequency ¹H and ¹³C nmr spectral studies of lyngbyatoxin A and a tetrahydro derivative obtained by catalytic hydrogenation of the toxin. The stereochemistry and absolute configuration of the lactam ring were deduced by comparing the optical rotations of lyngbyatoxin A,

TABLE 6

^{1}H nmr Data of Rearranged Oscillatoxin D in Benzene-d_6 (Incomplete)

δ in ppm[a]	Number of protons	Multiplicity; J in Hz	Assignment[b]
7.15	1	t; 8	20
7.09	1	dd; 2, 1	17
6.93	1	dt; 8, 1	21
6.85	1	ddd; 8, 2, 1	19
5.56	1	dd; 11, 1	8
5.30	1	dd; 11, 3	9
5.02	1	br t; 6	29
4.32	1	d; 12	31
4.06	1	t; 6.5	15
3.82	1	br s	
3.81	1	dd; 12, 5	30
3.15	1	br d[c]	11
3.14	3	s	OCH$_3$
2.50	1	br d; 18	28
2.09	1	dd; 18, 7	28
2.04	1	m	4
1.92	2	m	14, 10
1.72	1	m	14
1.61	1	m[d]	13
1.58	1	m	12
1.48	1	m[d]	13
1.14	1	dd; 15, 8	5
0.97	3	s	24
0.96	1	dd[e]	5
0.89	3	d; 7	26
0.85	3	d; 7	22
0.75	3	s	25
0.71	3	d; 7	23

[a] Relative to residual benzene-d_5 (δ 7.15) as internal standard.
[b] Verified by spin–spin decoupling experiments at 360 MHz.
[c] Obscured by OCH$_3$ signal.
[d] Tentative assignments.
[e] Obscured by C-24 methyl signal.

$[\alpha]_D$ $-171°$ (CHCl$_3$), and the closely related teleocidin B (**69**), and the CD curves of tetrahydrolyngbyatoxin A (**70**) and dihydroteleocidin B (**71**). The absolute stereochemistry of **69** and **71** had been established by an X-ray crystallographic study of dihydroteleocidin B monobromoacetate (**72**) (Sakabe *et al.*, 1966).

Examination of the 360-MHz ^{1}H nmr spectrum of lyngbyatoxin A in CDCl$_3$ (Fig. 1) shows the presence of small signals for another compound. This minor constituent could not be separated by gel filtration on

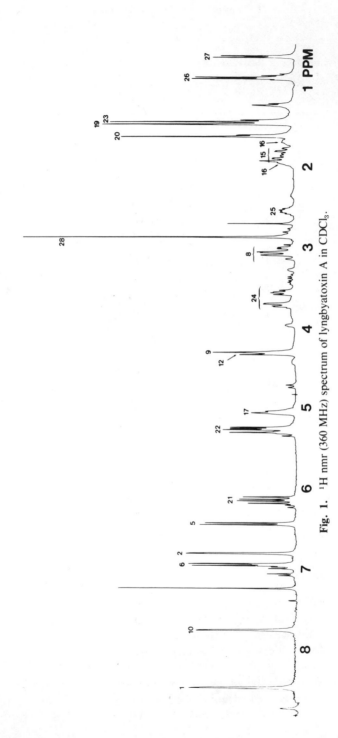

Fig. 1. ¹H nmr (360 MHz) spectrum of lyngbyatoxin A in CDCl₃.

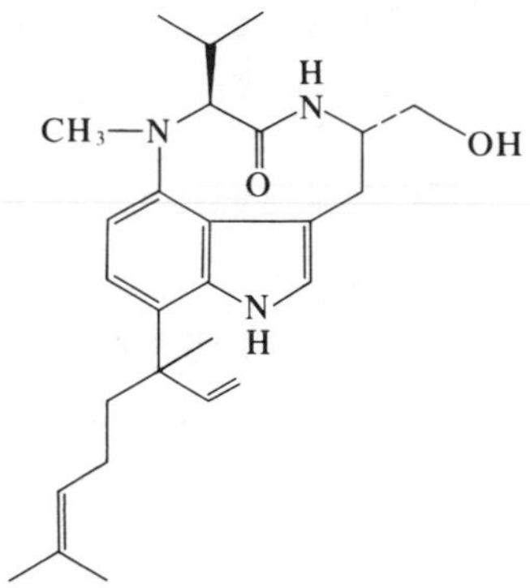

68

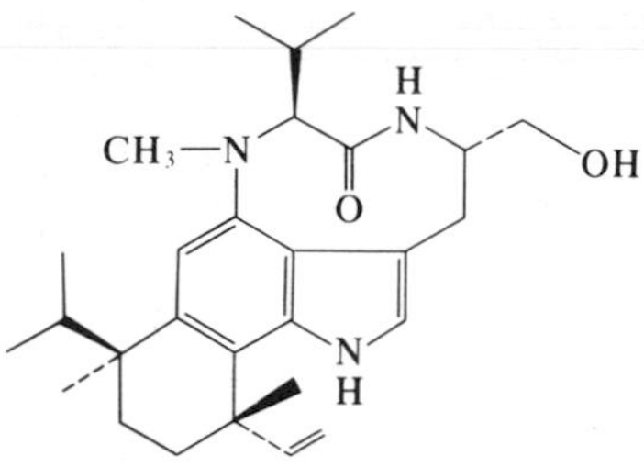

69

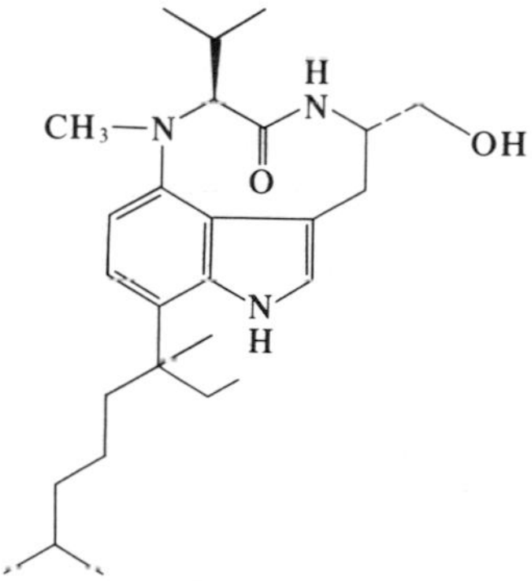

70

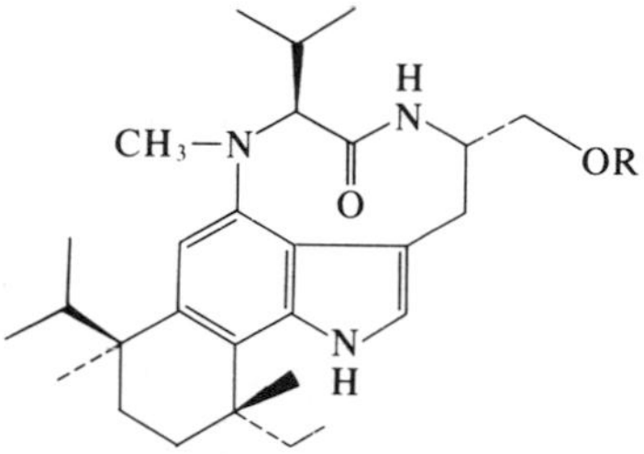

71　R = II
72　R = Ac

Sephadex LH-20 or reversed-phase liquid chromatography. Field desorption mass spectral analysis indicated that the minor constituent had the same molecular weight and elemental composition as lyngbyatoxin A. The chemical shifts of the smaller signals, e.g., the sharp doublets at 7.07 and 6.99 ppm for two vicinal aromatic protons, the broad singlet at 8.75 ppm for the indole NH, the doublet of doublets at 6.20 ppm for the vinyl methine proton, and the singlet at 2.70 ppm for the N-methyl group, are compatible with the placement of the linalyl group at C-5 instead of at C-7. The minor constituent, therefore, might have structure **73**.

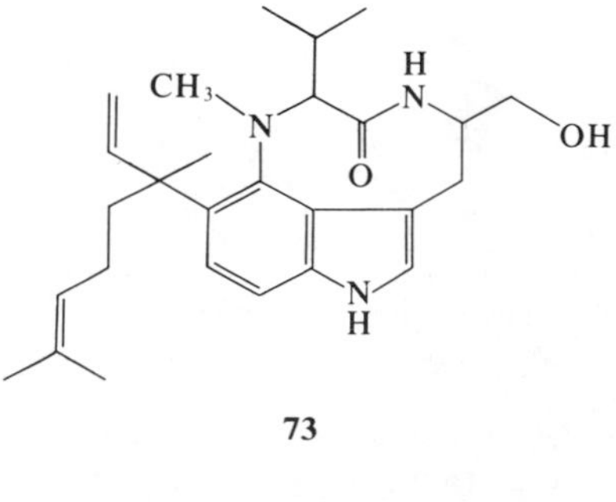

73

[1]H and ^{13}C nmr data for tetrahydrolyngbyatoxin A, which were not presented in Cardellina *et al.* (1979c), are given in Table 7.

4. Tolytoxins

J. S. Mynderse and R. E. Moore (unpublished work, 1978) found a toxic strain of *Tolypothrix conglutinata* var. *clorata* on the moist wall of a shed

TABLE 7

Nuclear Magnetic Resonance Data for Tetrahydrolyngbyatoxin A (70)

Carbon-13 δ in ppm[a] (multiplicity)	Assignment[b]	Proton δ in ppm (multiplicity; J in Hz)
174.6 (s)	11	
145.9 (s)	4	
137.3 (s)	7a	
123.9 (s)	7	
120.6 (d)	2	6.87 (s)[c]
121.9 (d)	6	6.86 (d; 8)
118.9 (s)	3 or 3a	
114.0 (s)	3 or 3a	
106.4 (d)	5	6.45 (d; 8)
71.0 (d)	12	4.35 (d; 12)
65.1 (t)	24	3.74 (dd; 12, 3); 3.58 (dd; 12, 9)
55.6 (d)	9	4.40 (br m)
42.1 (t)	15	1.79 (td; 12, 4); 1.65 (dq; 14, 7)
41.1 (s)	14	
39.6 (t)	17	1.06 (m)
33.9 (t)	8	3.15 (dd; 17, 2); 3.08 (dd; 17, 3)
33.9 (t)	21	1.94 (dq; 14, 7); 1.65 (dq; 14, 7)
32.9 (q)	28	2.89 (s)
28.4 (d)	25	2.57 (m)
27.6 (d)	18	1.43 (m)
24.4 (q)	23	1.39 (s)
22.7 (q)	19	0.76 (d; 7)
22.4 (q)	20	0.74 (d; 7)
22.0 (t)	16	1.18 (m); 0.88 (m)
21.6 (q)	26	0.91 (d; 6.5)
19.4 (q)	27	0.60 (d; 6.5)
8.9 (q)	22	0.63 (t; 7)
	1	8.21 (br s)
	1	7.70 (br s)[d]

[a] Relative to $CDCl_3$ (76.9 ppm) as internal standard.
[b] Based on proton single frequency off-resonance decoupling experiments at 90 MHz (carbon-13) and proton spin–spin decoupling experiments at 360 MHz.
[c] Becomes a sharper singlet when irradiated at 8.21 ppm.
[b] Becomes a sharper singlet when irradiated at 4.40 ppm.

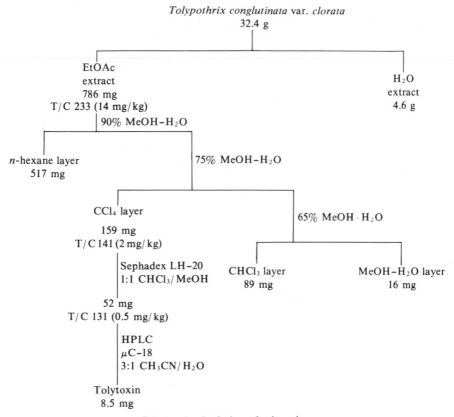

Scheme 4. Isolation of tolytoxin.

near the Cable Station, Fanning Island. The ethyl acetate extract of this alga, which showed good activity against P-388 lymphocytic mouse leukemia (Table 3) and inhibited the growth of some fungi and bacteria, was fractionated as outlined in Scheme 4. Tolytoxin, which appears to be the major anticancer compound in this alga, was isolated in about 0.03% yield and partially characterized.

The field desorption mass spectrum of tolytoxin suggests that the molecular weight is 1275.

The 220-MHz ^{1}H nmr spectrum in $CDCl_3$, which is complicated since the molecule exists in two conformations that interconvert slowly in all solvents at room temperature, is shown in Fig. 2. The portion of the molecule undergoing the slow conformational interconversion is shown in partial structures **74a** and **74b** in Scheme 5. The ^{1}H nmr data supporting the partial structures of the two conformers are summarized in Table 8.

Five *O*-methyl, one *N*-methyl, and six *C*-methyl groups are present in tolytoxin. One of the methoxyl groups (δ 3.63) is probably in a methyl

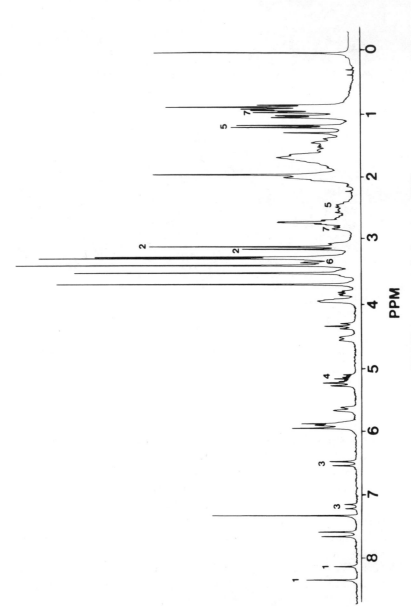

Fig. 2. ^{1}H nmr (220 MHz) spectrum of tolytoxin in CDCl$_3$.

TABLE 8

Some ^1H nmr Data of Tolytoxin in $CDCl_3$

Assignment[a]	δ in ppm (multiplicity; J in Hz)	
	Conformer 74a	Conformer 74b
1	8.08 (s)	8.29(s)
CH_3 on 2	3.08 (s)	3.04 (s)
3	7.14 (d; 15)	6.46 (d; 15)
4	5.16 (dd; 15, 8)	5.13 (dd; 15, 8)
5	2.39 (m)	2.39 (m)
CH_3 on 5	1.15 (d; 7)	1.15 (d; 7)
6	3.32 (dd; 10, 2)	3.32 (dd; 10, 2)
7	2.68 (dq; 10, 7)	2.68 (dq; 10, 7)
CH_3 on 7	0.90 (d; 7)	0.92 (d; 7)

[a] Supported by spin–spin decoupling experiments at 220 and 360 MHz.

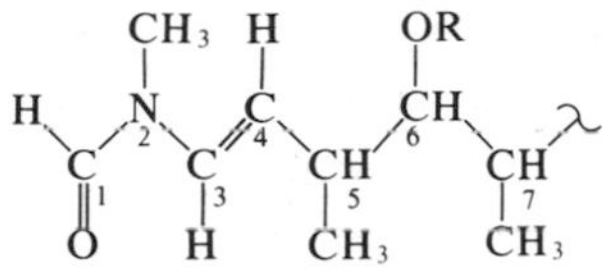

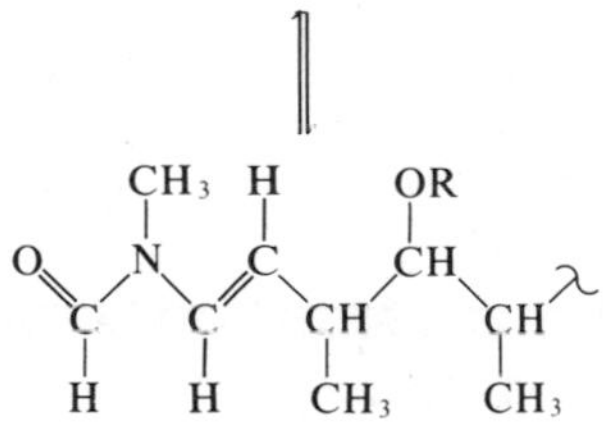

Scheme 5.

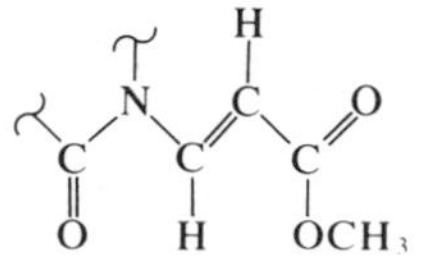

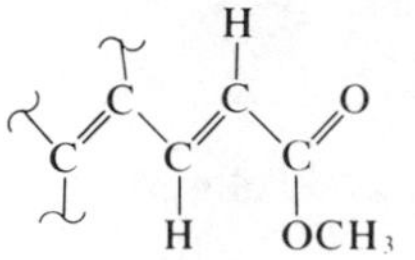

ester functionality of either a *trans*-β-amidoacrylate (**75**) or a γ-substituted α,β,γ,δ-dienoate (**76**) system. The uv spectrum, which shows a single absorption maximum at 258 nm, and the infrared spectrum, which exhibits a shoulder at 1710 cm^{-1} and bands at 1685 and 1660 cm^{-1}, agree with either structure. The ^1H nmr spectrum of tolytoxin shows doublets at 5.87 and 7.57 ppm (J = 16 Hz), and the chemical shifts are comparable with those of the α and β protons of appropriate model compounds such as methyl E-β-acetamidoacrylate and methyl sorbate. The coupling constant of 16 Hz, however, favors structure **76**. There is at least one other tolytoxin-related compound in *T. conglutinata* var. *clorata* from Fanning Island.

IV. FATTY ACIDS, ACETOGENINS, AND RELATED COMPOUNDS

The fatty acid compositions of blue-green algae are different from those of bacteria. Polyunsaturated fatty acids are absent in bacteria and the major monounsaturated acid is the Δ-11 C_{18} vaccenic acid. In contrast, the major monoenoic acid found in blue-green algae is the Δ-9 C_{18} oleic acid, and polyunsaturated fatty acids are generally present in the morphologically more complex forms (Holton *et al.*, 1968). A specific comparison can be made between *Oscillatoria* and its colorless bacterial counterpart *Beggiatoa*, a gliding organism although not a eubacterium. Erwin and Bloch (1963) report the presence of an unspecified monounsaturated fatty acid in *Beggiatoa*, but doubly and triply unsaturated fatty acids were found in all *Oscillatoria* species examined.

The most common fatty acids in blue-green algae are C_{16} and C_{18}. Unusual fatty acids are sometimes present as major constituents. Malyngic acid, for example, a major fatty acid in shallow-water and deep-water varieties of *Lyngbya majuscula*, is 9(S),12(R),13(S)-trihydroxyoctadeca-10(E),15(Z)-dienoic acid (**77**). The structure of malyngic acid was determined from chemical and spectral data and by its conversion to 9(S), 12(R), 13(S)-trihydroxystearic acid and degradation to 2-deoxy-D-ribitol Moore, 1980). 9-Methoxy-9-methylhexadeca-4(E),8(E)-dienoic acid (**78**) is present in moderate amount in a deep-water variety of *L. majuscula* from Enewetak (Loui and Moore, 1979).

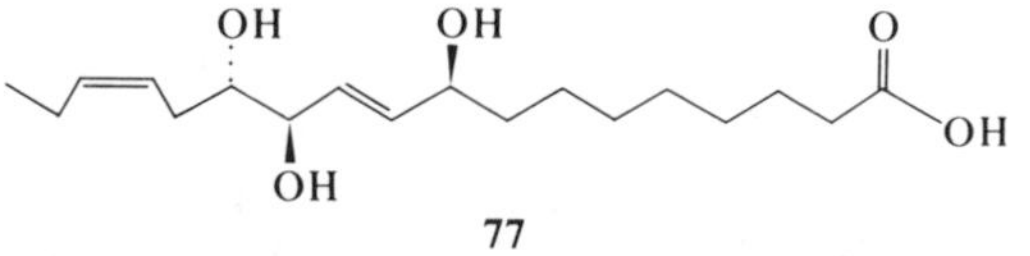

77

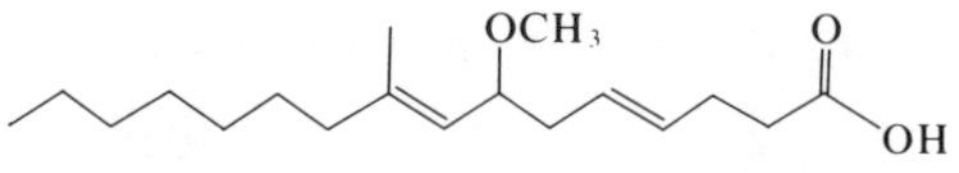

78

C_{14} fatty acids and related compounds are frequently present in marine Oscillatoriaceae. 7(S)-Methoxytetradec-4(E)-enoic acid (**79**) is a major fatty acid in all shallow-water varieties of *L. majuscula* (Cardellina *et al.*, 1978). Amides of **79** are also present in shallow-water varieties of *L. majuscula* (see Section VII). This fatty acid was not detected in *L. semiplena* and several species of *Schizothrix, Oscillatoria, Symploca, Phormidium*, and *Spirulina*, but **79** was found to be a major constituent of a cyanophyte tentatively identified as a *Crinalium* sp. 1-Chlorotridec-1(*E*)-ene-6(*R*),8(*R*)-diol (**80**) is a major constituent in a toxic 1:1 mixture of blue-green algae, tentatively identified as *Schizothrix calcicola* and *Oscillatoria nigroviridis*, from Enewetak Atoll (Mynderse and Moore, 1978c).

79   80

Malyngolide (**81**), a lipophilic constituent of *L. majuscula,* appears to be derived from a C_{14} fatty acid. This δ-lactone is responsible for the antibacterial activity reported for crude extracts of Hawaiian *L. majuscula* (Starr *et al.*, 1962). Malyngolide, like the crude extract of Hawaiian *L. majuscula,* shows good activity against *Mycobacterium smegmatis* and *Streptococcus pyogenes,* but is inactive toward *Escherichia coli* and *Pseudomonas aeruginosa.* The skeletal structure and relative stereochemistry of malyngolide were established by spectral analysis, chemical degradation, and a lanthanide-induced proton chemical shift study (Cardellina *et al.*, 1979d). The absolute configuration was determined by a CD study and by converting malyngolide to levorotatory methyl 2-methyl-5-oxotetradecanoate which had optical properties that were comparable with those of methyl 2(*R*)-methyl-5-oxoheptanoate. X-Ray crystallographic analysis of the levorotatory saponification product, (2*R*,5*S*)-5-hydroxy-5-(hydroxymethyl)-2-methyltetradecanoic acid, also verified the absolute stereochemistry.

81

Extracts of *L. majuscula* from Puerto Rico are reported to show good activity against *E. coli,* but only a trace of or no activity against *M. smegmatis* (Burkholder *et al.*, 1960). The active compound, however, has not been isolated and identified.

4(*S**), 6(*S**), 8(*S**), 10(*S**), 12(*R**), 14(*R**), 16(*R**), 18(*R**), 20(*R**)-Nonamethoxy-1-pentacosene (**82**) and smaller amounts of the isotactic (i.e., all methoxyls on same side of carbon chain) homologs, 4,6,8,10,12,14,-

16,18,20,22- decamethoxy-1-haptacosene **(83)** and 4,6,8,10,12,14,16,18-octamethoxy-1-tricosene **(84)**, are novel lipophilic constituents of the

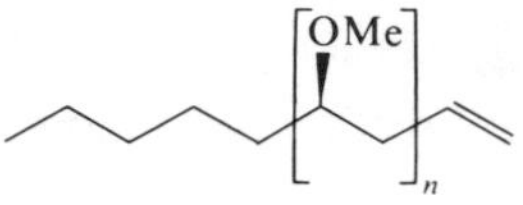

82 $n = 9$
83 $n = 10$
84 $n = 8$

toxic blue-green alga *Tolypothrix conglutinata* var. *clorata* from Fanning Island (Mynderse and Moore, 1979). The structures were assigned from mass and nmr spectral data.

Poly-β-hydroxybutyric acid (PHB) is a naturally occurring polymer which is found in a large number of photosynthetic and nonphotosynthetic bacteria. The polyester was first discovered in a bacterium by Lemoigne (1925, 1927); Carr (1966) demonstrated that it also is present in the blue-green alga *Chlorogloea fritschii*. J. S. Mynderse and R. E. Moore (unpublished work, 1979) have found it in an *Oscillatoria* sp. from Fanning Island. It occurs as crystalline hydrophobic granules in the cytoplasm of bacteria (Merrick and Doudoroff, 1964) and blue-green algae (Jensen and Sicko, 1971) and has clearly been shown to be a carbon and energy reserve in prokaryotes (Dawes and Gibbons, 1964), a role that is similar to that of starch in eukaryotic plants. PHB is optically active (Baptist and Weber, 1964), and chemical and spectral studies as well as synthetic studies (Shelton *et al.*, 1971a,b) support a linear head-to-tail structure **(85)** based on levorotatory R-β-hydroxybutyric acid. X-Ray

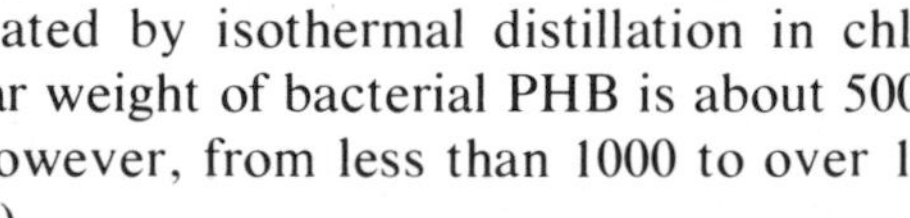

85

crystallographic studies suggest a right-handed 2_1 helical conformation for the polymer (Cornibert and Marchessault, 1972). Williamson and Wilkinson (1958) estimated by isothermal distillation in chloroform that the average molecular weight of bacterial PHB is about 5000. The molecular weight ranges, however, from less than 1000 to over 100,000 (Marchessault *et al.*, 1970).

V. VOLATILE CONSTITUENTS

Gaseous and volatile substances are continually being emitted into the atmosphere by marine algae. Only recently have scientists begun to study the transfer processes quantitatively. The amount must be large when one considers the entire algal biomass of the ocean, most of which is planktonic. Marine algae, for example, contribute appreciably to the natural pool of carbon monoxide. In blue-green algae CO is a by-product of phycocyanobilin formation (Troxler and Dokos, 1973), and this has been clearly demonstrated by the incorporation of δ-[5-^{14}C]aminolevulinic acid into CO and the pigment (Troxler *et al.*, 1970). Similarly, CO is produced concomitantly with phycoerythrobilin formation.

Some blue-green algae possess gas vacuoles which presumably allow the cyanophytes to travel between the water surface and the ocean or lake bottoms where the nutrients are located. The gas vacuoles diminish in size at high light intensities, and the cells lose their ability to float. The biochemistry associated with these gas vacuoles is unknown. Gas vacuoles are also found in certain bacteria but nowhere else.

Persistent earthy and musty odors in raw water are frequently associated with the presence of blue-green algae. The major causative agent of these unpleasant odors appears to be geosmin, which was isolated from *Symploca muscorum* and *Oscillatoria tenuis* (Safferman *et al.*, 1967; Medsker *et al.*, 1968). The gross structure of geosmin was deduced from spectral and chemical degradation data and shown to be *trans*-1,10-dimethyl-*trans*-9-decalol (**86**) by direct comparison with a synthetic

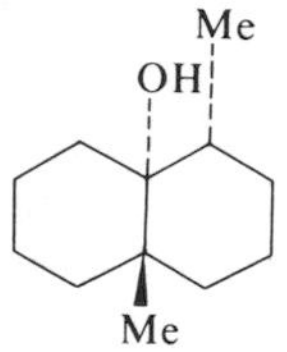

86

sample (Gerber, 1968). The absolute configuration was not determined. Biogenetically, geosmin appears to be a sesquiterpene that has lost an isopropyl group.

Odoriferous organosulfur compounds have been identified in cultures of blue-green algae. Associated bacteria, however, may play a role in the production of these compounds. *Microcystis flos-aquae* produces isopropyl mercaptan during periods of active growth (Jenkins *et al.*, 1967).

A tobacco-like odorous compound has been isolated from a bloom of *Microcystis wesenbergii* and from a laboratory culture of *M. aeruginosa* and identified as β-cyclocitral (**87**) (Jüttner, 1976). No other monoterpenes

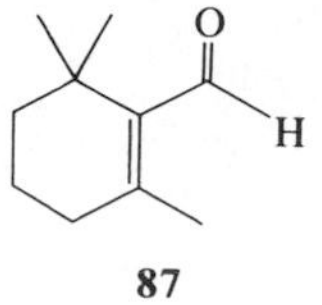

87

were detected. β-Cyclocitral is probably formed by degradation of carotenes in the plant rather than by direct monoterpene biosynthesis.

VI. ISOPRENOIDS AND STEROLS

With the exception of β-cyclocitral, monoterpenes have so far not been detected in blue-green algae. Sesquiterpenes and diterpenes have also not been found.

Sterols were once believed to be absent in blue-green algae (Levin and Bloch, 1964). In 1968, however, Reitz and Hamilton reported the isolation of cholesterol and its 24-ethyl derivative from the saponified extracts of *Anacystis nidulans* and *Fremyella diplosiphon,* and de Souza and Nes (1968) isolated sterols from the acetone extract of *Phormidium luridum,* the major ones being 24-ethyl-Δ^7-cholesterol and 24-ethyl-$\Delta^{7,22}$-cholestadienol with smaller amounts of cholesterol, 24-ethyl-$\Delta^{5,7,22}$-cholestatrienol, 24-ethyl-ethyl-$\Delta^{5,22}$cholestadienol, 24-ethyl-$\Delta^{5,7}$-cholestadienol, and 24-ethyl-Δ^5-cholesterol. In *Chlorogloea fritschii,* sterol quantity and composition are dependent on the nature of the nitrogen source present in the growth medium. N. G. Carr and L. J. Goad, as cited by Nichols (1973), found that the yield of sterols was 0.13% of the cellular dry weight and that the sterols were composed mainly of stigmasterol with smaller amounts of cholesterol and an unidentified sterol when the nitrogen source was ammonium ion; when the nitrogen source was nitrate, the yield of sterol was much smaller (0.01%), but only cholesterol and the unknown sterol were produced. Similarly, bacteria were once thought to lack sterols, but these prokaryotes also produce sterols (Schubert *et al.,* 1968).

Squalene has been isolated from prokaryotic organisms (Bird *et al.,* 1971a) as have the squalene-cyclized triterpenes, hop-22(29)-ene (**88,** diploptene) and hop-17-ene (De Rosa *et al.,* 1971). Hop-22(29)-ene is present in the blue-green algae *Nostoc* sp., *Lyngbya estuarii,* and *Chroococcus turgidus* (Bird *et al.,* 1971b).

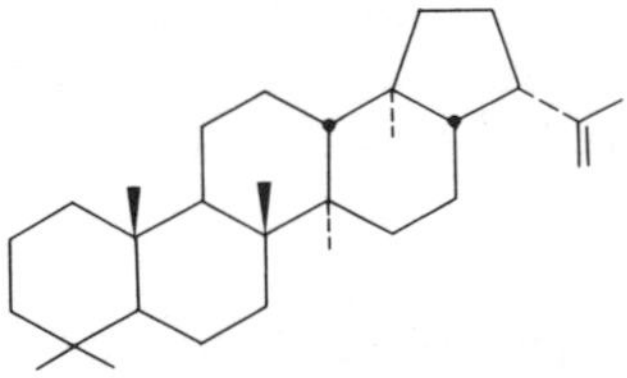

88

Novel C_{35} derivatives of hopane substituted with a five-carbon chain bearing hydroxyl groups are present in prokaryotes. These complex lipids, the parent hydrocarbon of which is called bacteriohopane, were first isolated from the bacterium *Acetobacter xylinum* (Förster *et al.*, 1973). The structure of the major bacteriohopane, a tetraol (**89**), was believed to have the C_5 tetraol unit attached to C-22 of the hopane moiety on the basis of mass spectral evidence. It was subsequently shown from nmr data and synthesis of a degradation product that the C_5 tetraol unit was attached to C-29 (Rohmer and Ourisson, 1976). The tetraol was furthermore shown to be the 22*R* epimer. The tetraol and two pentahydroxybacteriohopanes (**90** and **91**) have been found in the blue-green alga *Nostoc muscorum* along with a Δ^6-unsaturated bacteriohopane (**92**). Similar compounds were also detected in *Anabaena variabilis*.

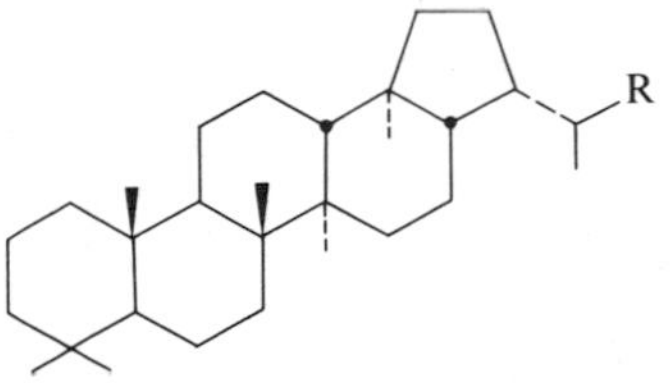

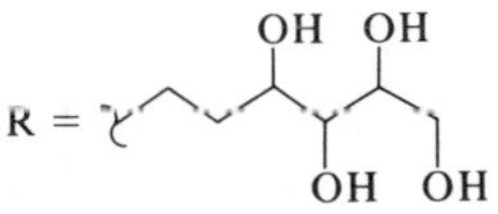

89 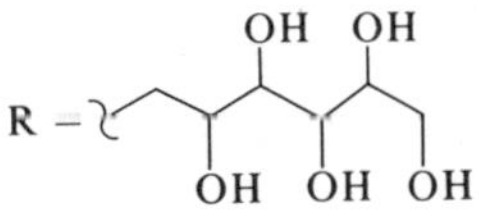   90 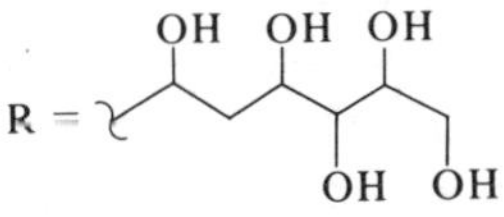   91

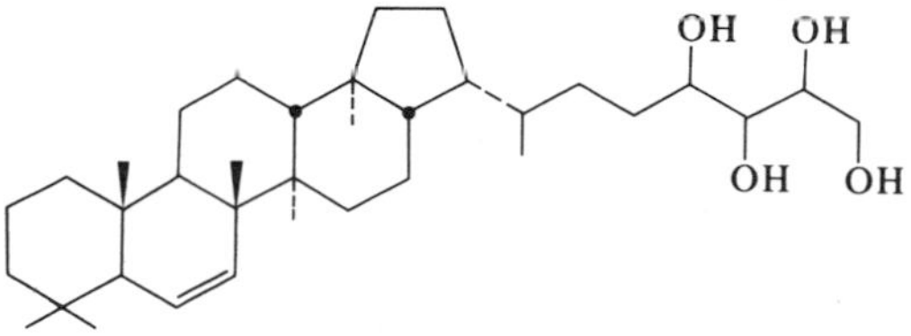

92

The biosynthesis of the bacteriohopanes probably proceeds by the condensation of a C_5 unit and the preformed pentacyclic triterpene system. The isolation of hopenes and 22-hydroxyhopane (Förster *et al.*, 1973) from prokaryotes supports the proposed biogenesis. Bacteriohopanes may have been the precursors of hopane geolipids having more than 30 carbon atoms.

VII. NITROGENOUS COMPOUNDS

Nitrogen fixation is the enzymatic conversion of atmospheric nitrogen to ammonia. Only prokaryotes are capable of carrying out this reduction. The fixation of nitrogen by blue-green algae was first unequivocally demonstrated by Drewes in 1928. Even though this demonstration occurred 35 years after nitrogen fixation was first discovered in a heterotrophic bacterium by Winogradsky, more genera of blue-green algae are now known to fix nitrogen than there are heterotrophic and photosynthetic bacteria combined (Stewart, 1973). Heterocystous species are the most common nitrogen-fixing blue-green algae, but a few unicellular and nonheterocystous filamentous forms are also able to fix nitrogen.

In contrast to the situation in eukaryotic algae, nitrogen-containing compounds are common in blue-green algae. Relatively few compounds have been isolated and characterized, however, mainly because algal samples of sufficient quantity and purity are often difficult to obtain in the field. Culturing is frequently needed. Alkaloids are present in blue-green algae (see Section III). Anatoxin-a (7) and saxitoxin (28) are the first examples of alkaloids from freshwater species and lyngbyatoxin A (68) represents the first alkaloid to be isolated from a marine species. Although the picture is fragmentary at present, it may be predicted that many other types of alkaloids will be discovered as research grows and intensifies on this interesting and important group of organisms.

The Moore group has found several novel nitrogenous compounds in species of marine blue-green algae. In the following subsections the chemical and biological significance of these nitrogen-containing compounds is presented.

A. Malyngamides

Shallow-water varieties of *Lyngbya majuscula* contain amides of a novel fatty acid, 7(S)-methoxytetradec-4(E)-enoic acid (79); deep-water varieties of *L. majuscula,* on the other hand, contain amides of 7-methoxy-9-methylhexadec-4(E)-enoic acid (93) (Cardellina *et al.*, 1978). Both types of amides are called malyngamides.

93

Malyngamide A (94), a constituent of several shallow-water varieties of *L. majuscula* from Hawaii, is a chlorine-containing amide of 79 (Cardellina *et al.*, 1979b). Its molecular formula, $C_{29}H_{45}ClN_2O_6$, was determined

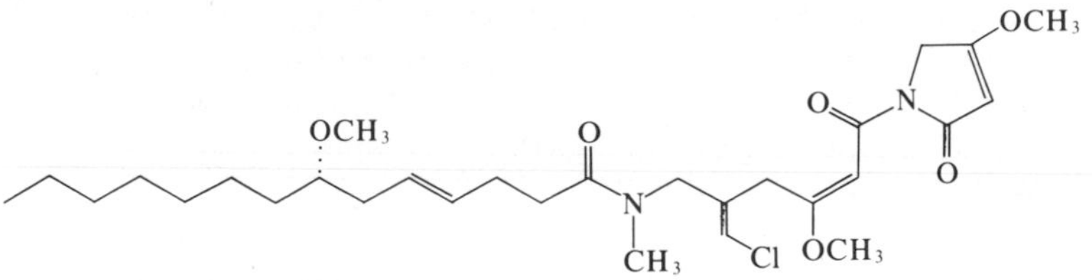

94

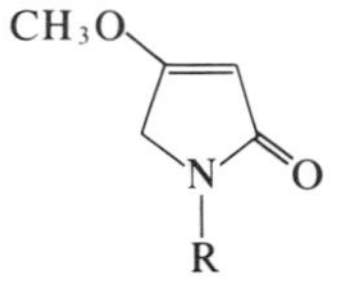

95 R = H
96 R = COCH₃

by mass spectrometry and its structure was elucidated from chemical and spectral data. Hydrolysis of **94** led to 4-methoxy-Δ³-pyrrolin-2-one (**95**) in addition to **79**. ¹H nmr analysis indicated that the pyrrolinone moiety was N-acylated, since the ¹H chemical shifts for this unit were essentially identical with those for N-acetyl-4-methoxy-Δ³-pyrrolin-2-one (**96**). Compounds **95** and **96** were also constituents of the alga. The geometry of the alkenyl chloride functionality was established as *E* from an NOE experi-

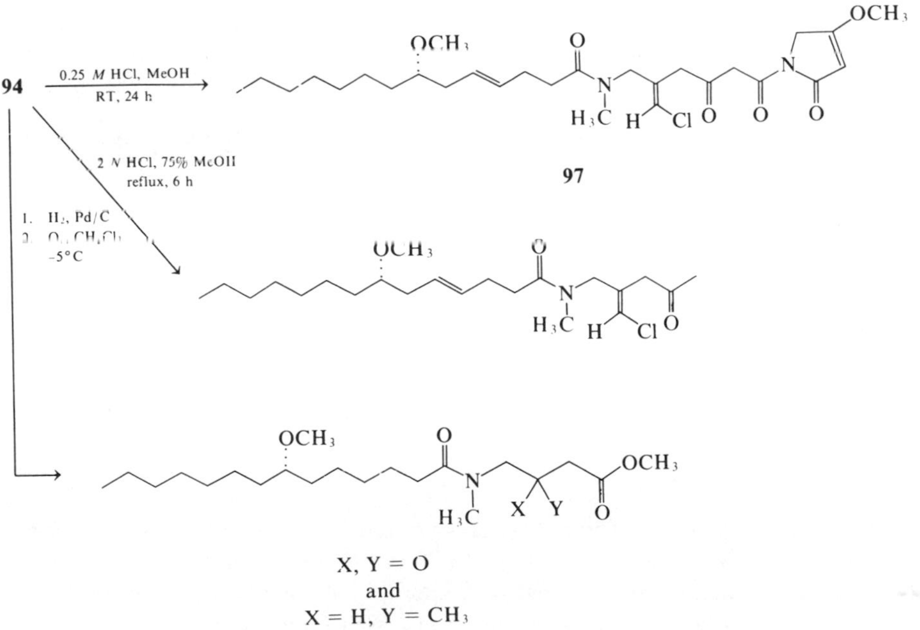

Scheme 6. Chemical degradations and transformations of malyngamide A (**94**).

ment. Comparison of uv and ^{1}H nmr spectral data of malyngamide A and of the β-ketoamide **97** from mild acid hydrolysis (Scheme 6) with those of the sponge metabolite dysidin (**98**) and its acid hydrolysis product **99**, indicated the presence of an acyclic β-methoxyenamide system in malyngamide A, which had an *E* configuration. Structure **94** was consistent with other chemical degradations and transformations outlined in Scheme 6.

Dysidin was isolated from an Australian variety of the sponge *Dysidea herbacea*. Three classes of compounds have been isolated from *D. herbacea*, viz., sesquiterpenes (Kazlauskas *et al.*, 1978a), polybrominated diphenyl ethers (Sharma and Vig, 1972; Sharma *et al.*, 1969), and polychlorinated amino acid-derived metabolites such as dysidin (**98**) (Hofheinz and Oberhänsli, 1977), dysidenin (**100**) (Kazlauskas *et al.*, 1977), isodysidenin (**101**) (Charles *et al.*, 1978), and the trichloroleucine-derived diketopiperazine **102** (Kazlauskas *et al.*, 1978b). Interestingly,

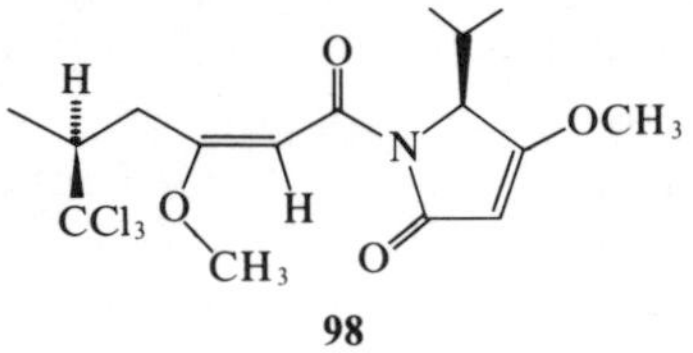

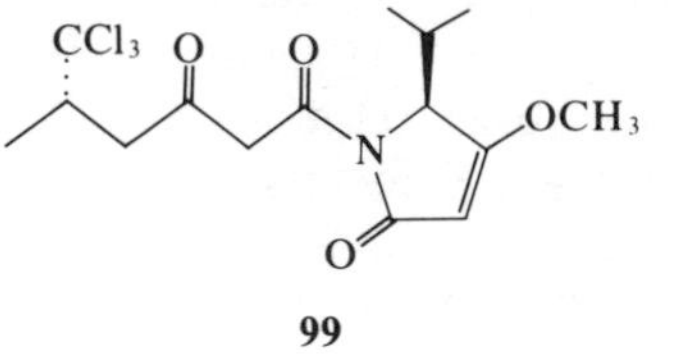

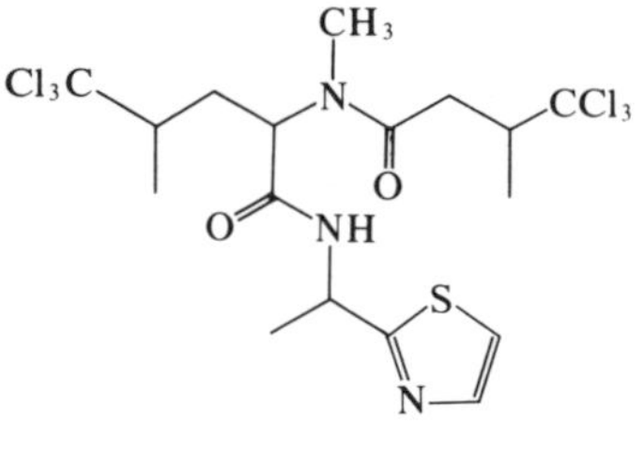

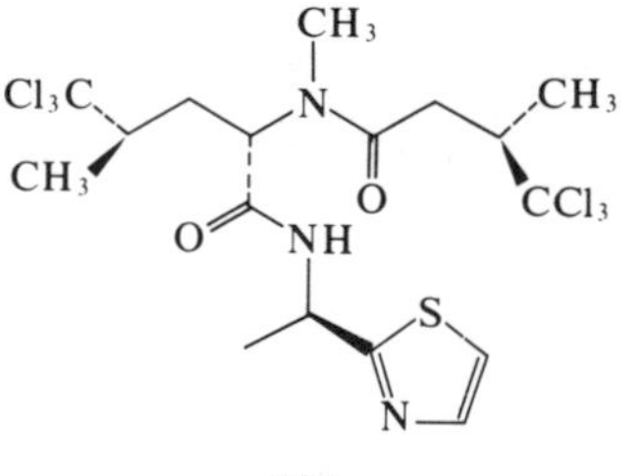

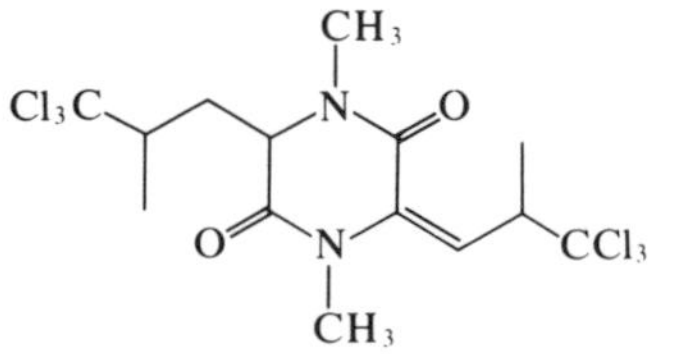

blue-green algae are frequently associated symbiotically with *D. herbacea;* e.g., *Phormidium spongeliae* has been identified as a symbiont (Charles *et al.*, 1978). At least two species of dominant symbiotic blue-green algae, which often constitute over 50% of the cellular material, have been observed in *D. herbacea* (R. J. Wells, unpublished work, 1979). Sesquiterpenes are the major secondary metabolites in those specimens of *D. herbacea* that lack symbiotic blue-green algae and bacteria. Symbiotic prokaryotes, however, are present in varieties of *D. herbacea* that produce polybromodiphenyl ethers and polychlorinated amino-acid-derived metabolites. In the varieties that elaborate polybromodiphenyl ethers, sesquiterpenes have never been detected, but sesquiterpenes are found in varieties containing polychlorinated amino acid-derived metabolites. Metabolites of the sponge may therefore be acting as carbon sources, which are utilized effectively by certain dominant symbionts to produce polybromodiphenyl ethers instead of sesquiterpenes, but ineffectively by other symbionts to produce polychlorinated amino acid-derived metabolites along with sesquiterpenes. The structural similarity of malyngamide Λ (**94**) and dysidin (**98**) provides some circumstantial evidence that dysidin may be a metabolite of a symbiotic blue-green alga in *D. herbacea*. The finding of *Lyngbya majuscula* var. *spongophila* in an Indonesian sponge (Drouet, 1968) provides further proof.

Malyngamide B, which was isolated from a Hawaiian shallow-water variety of *L. majuscula* (Cardellina *et al.*, 1978), is also a chlorine-containing amide of **79**. Its molecular formula was shown to be $C_{28}H_{45}ClN_2O_6$ by mass spectrometry. Inspection of its ^1H nmr spectrum indicated that the *E* alkenyl chloride and β-methoxyenamide groups in **94** were also present in malyngamide B, but that the *N*-acylated 4-methoxy-Δ^3-pyrrolin-2-one unit was missing. Malyngamide B, which has one less methoxy group than **94,** most probably has structure **103**. Mild acid hydrolysis of malyngamide B leads to a β-ketoamide, presumably **104.**

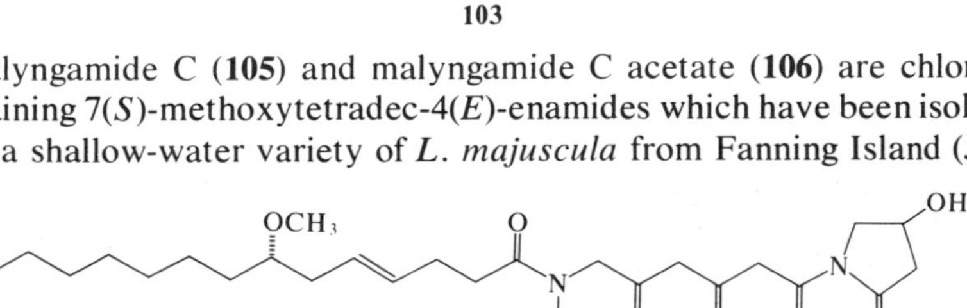

103

Malyngamide C (**105**) and malyngamide C acetate (**106**) are chlorine-containing 7(*S*)-methoxytetradec-4(*E*)-enamides which have been isolated from a shallow-water variety of *L. majuscula* from Fanning Island (J. S.

104

Mynderse and R. E. Moore, unpublished work, 1978). Mass spectrometry shows that the molecular composition of malyngamide C is $C_{24}H_{38}ClNO_5$ (Cardellina *et al.*, 1978). The structure of malyngamide C was determined on the basis of chemical and spectral data from malyngamide C and several derivatives. Treatment of malyngamide C with hydrazine led to the tetrahydrocinnoline **107**. Chromous sulfate reduction of **105** gave an α,β-unsaturated ketone **108** which on standing in chloroform solution was slowly converted to the indolic compound **109**. In the presence of dilute methanolic hydrochloric acid, malyngamide C is transformed into an isomer proposed to have sturcture **110**. The stereochemistry of malyn-gamide C and its derivatives have not been rigorously established. Malyn-gamide C is structurally similar to stylocheilamide (**111**) and deacetoxy-stylocheilamide (**112**), two nontoxic lipids from the sea hare *Stylocheilus*

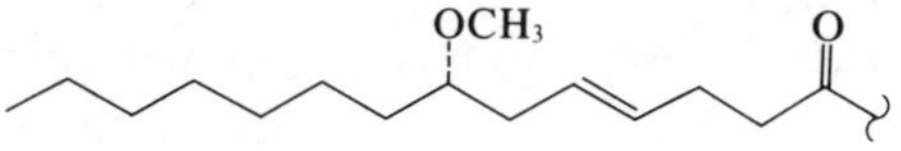

R_1

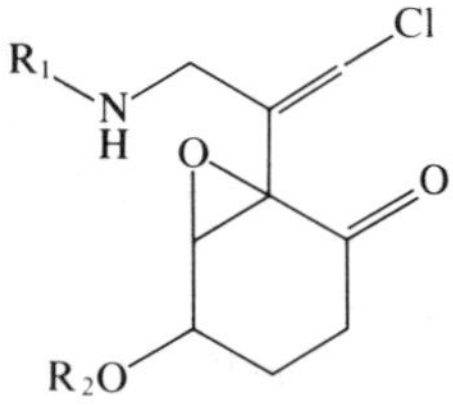

105 $R_2 = H$
106 $R_2 = Ac$

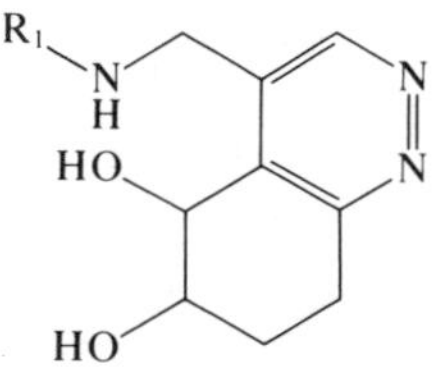

107

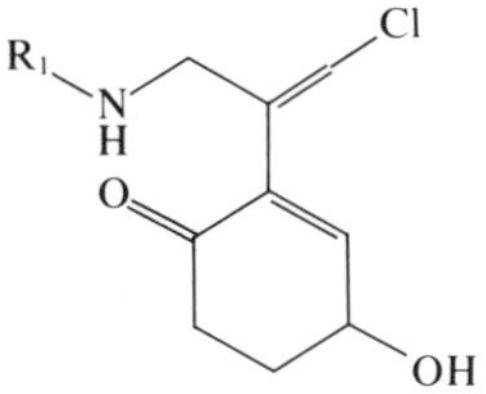

108

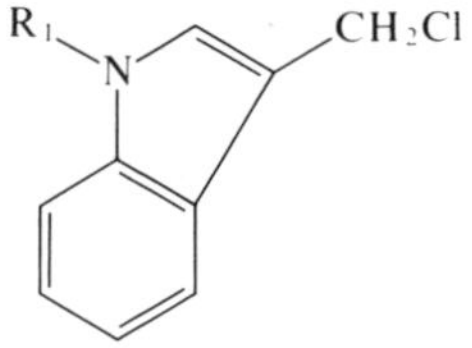

109

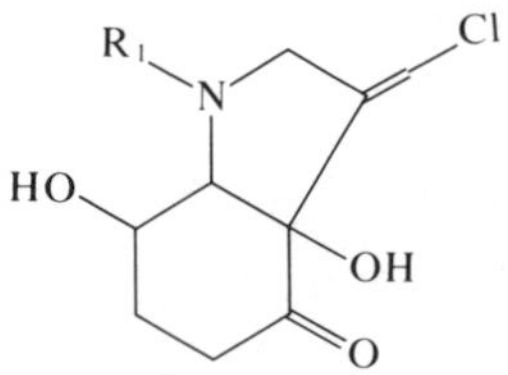

110

longicauda (Rose *et al.*, 1978). Compounds **111** and **112** obviously have a dietary origin and are probably constituents of *L. majuscula*.

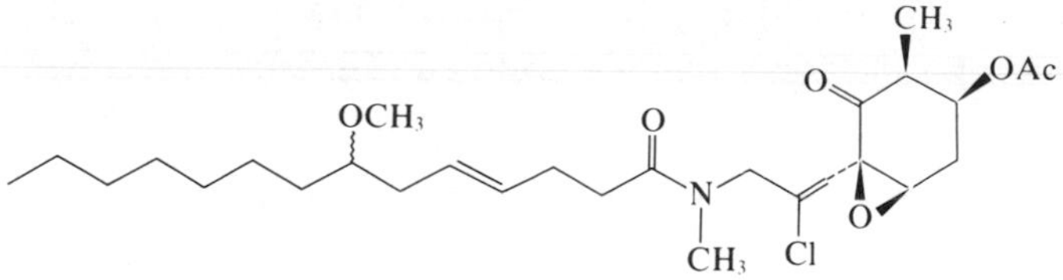

111

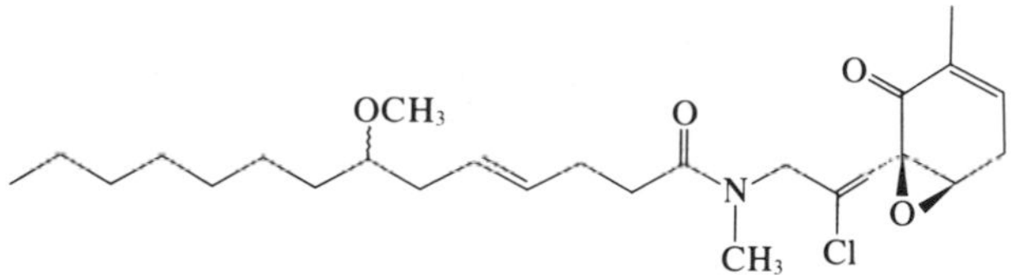

112

Malyngamides D and E are two closely related 7-methoxy-9-methylhexadec-4(*E*)-enamides that have been isolated from a deep-water variety of *L. majuscula* found on the pinnacles in Enewetak lagoon (Mynderse and Moore, 1978b). Detailed spectral analysis, in particular ^{1}H and ^{13}C nmr, and chemical degradation show that malyngamides D and E have the gross structures **113** and **114**, respectively. Malyngamides D and E produced the same diacetate **115** on acetylation. Acid hydrolysis of **114**

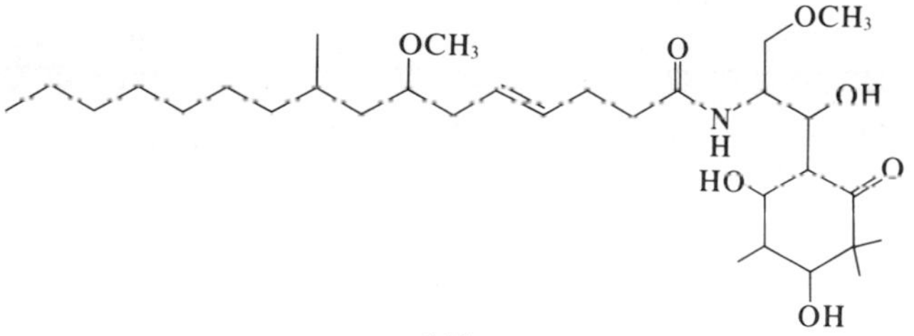

113

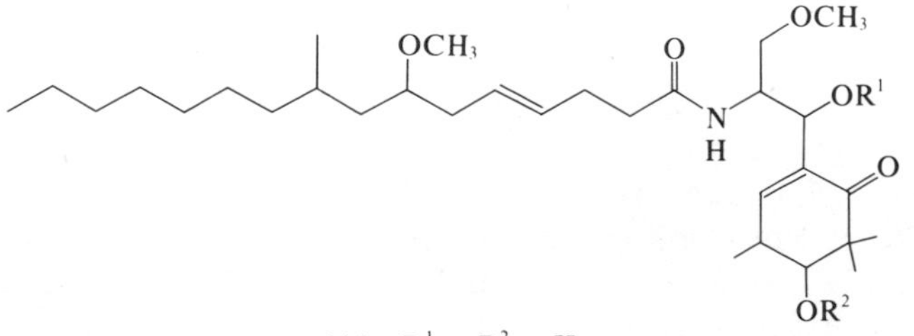

114 $R^1 = R^2 = H$
115 $R^1 = R^2 = Ac$

yielded 7-methoxy-9-methylhexadec-4(*E*)-enoic acid (**93**). The placement
of the methyl at C-9 was determined by a lanthanide-induced shift study of
aldehyde **116**, obtained by ozonolysis of **113** and **114**. The relative
stereochemistry of **116** was not determined. The ring stereochemistries of
113 and **114** have been defined as **113a** and **114a**, respectively, from nmr
and chemical reactivity data.

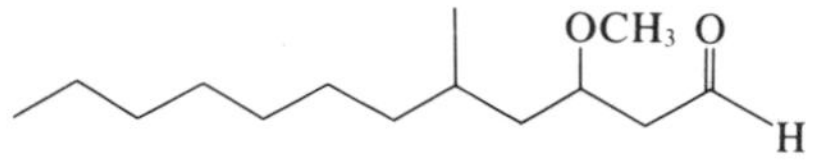

116

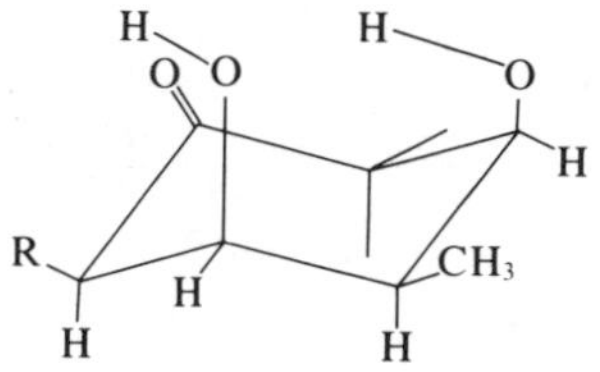

113a 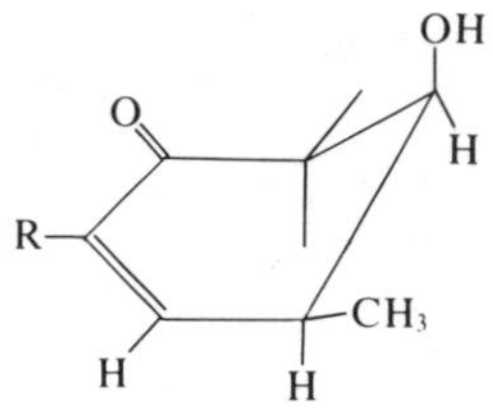   114a

B. Pukeleimides

Minor amounts of seven unusual pyrrolic compounds (pukeleimides,
after the Hawaiian word "pukele" which means "to gather thickly in the
water" such as an algal bloom), which appear to be related to malyngamide A (**94**), were found in a shallow-water variety of *L. majuscula* at
Kahala Beach, Oahu. The structure of pukeleimide C (**117**) was solved by
X-ray analysis (Simmons *et al.*, 1979), and the structures of pukeleimides
A, B, D, E, F, and G (**118–123**, respectively) were determined from
spectral data.

C. Majusculamides

Majusculamides A and B are major lipophilic constituents in several
shallow-water varieties of *L. majuscula* found in Hawaii and Enewetak
(Marner *et al.*, 1977). Detailed spectral analysis, chemical degradation,
and X-ray crystallographic studies showed that majusculamide B is *N*-
[(2*S*)-2-methyl-3-oxodecanoyl]-D-*N*,*O*-dimethyltyrosyl-L-*N*-methylvalin-
amide (**124**). Majusculamide A is the (2*R*)-2-methyl-3-oxodecanoyl epimer
(**125**). The chemistry of majusculamides A and B has been previously
reviewed in this series (Moore, 1978).

Majusculamide C is a major lipophilic constituent in several deep-water
varieties of *L. majuscula* found in Enewetak (D. C. Carter, J. S. Myn-

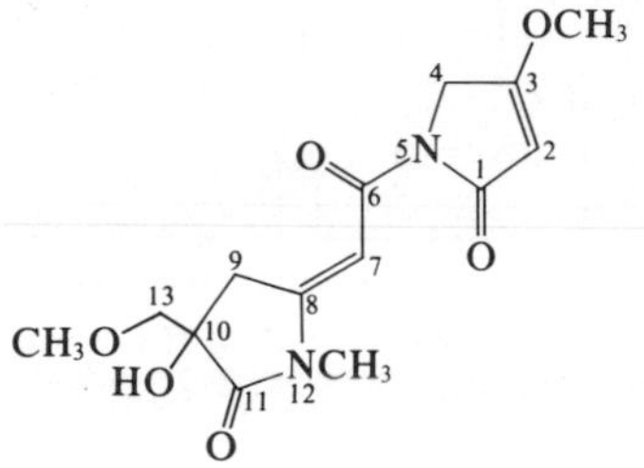

117

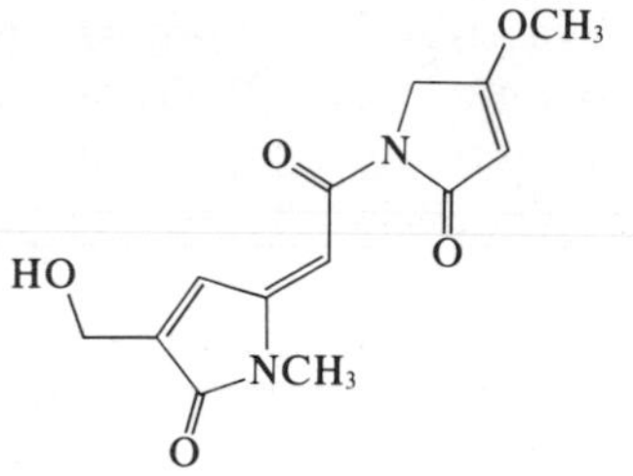

118

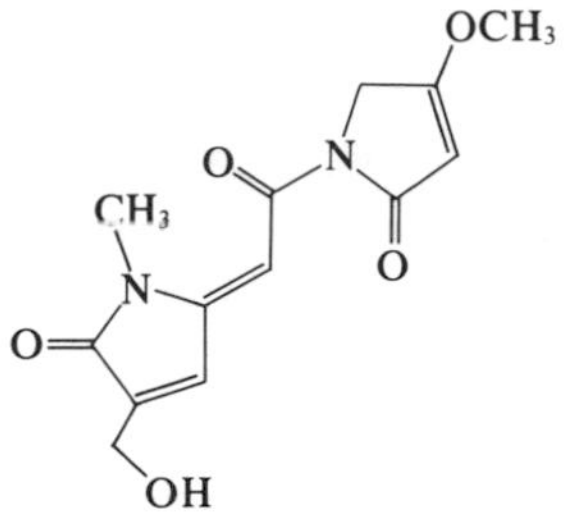

119

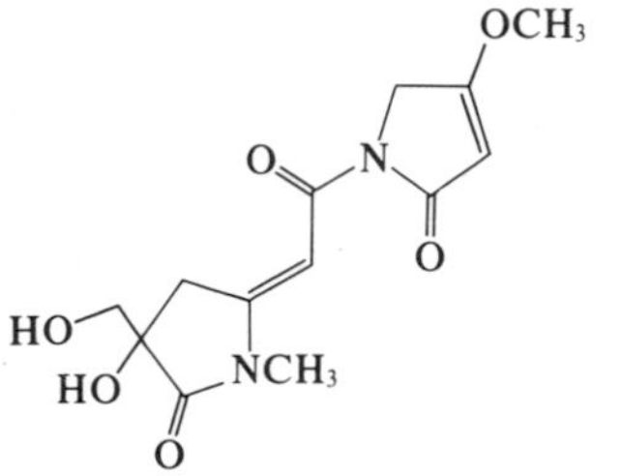

120

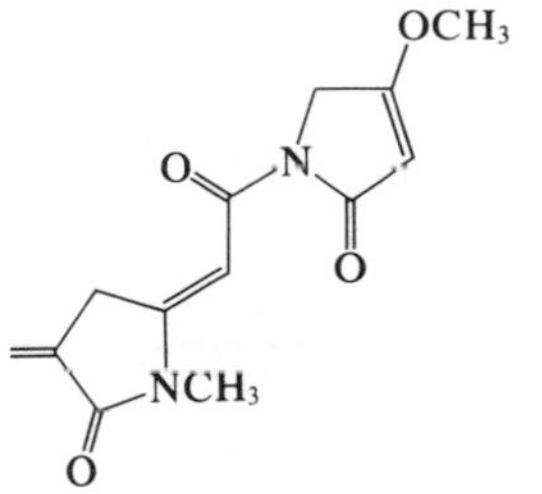

121

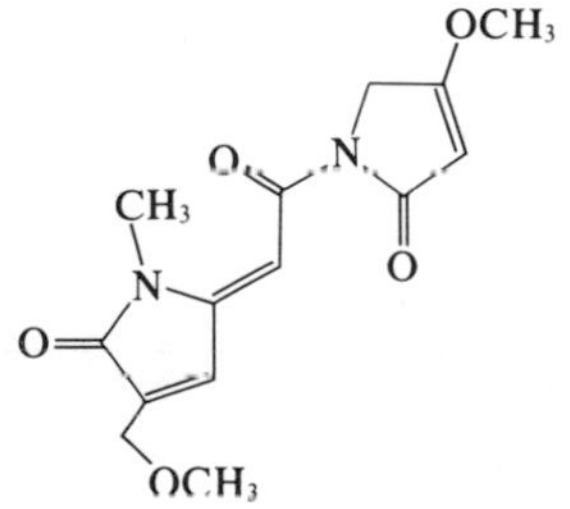

122

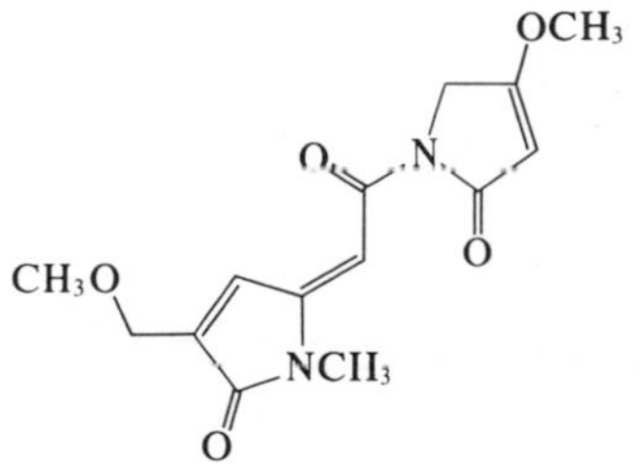

123

derse, and R. E. Moore, unpublished work, 1980). ^{1}H and ^{13}C nmr and mass spectral analysis indicates that the molecular composition is $C_{50}H_{80}N_8O_{12}$. Nuclear magnetic resonance analysis also shows that two glycyl units, two alanyl units, one N-methylvalyl unit, and one N,O-

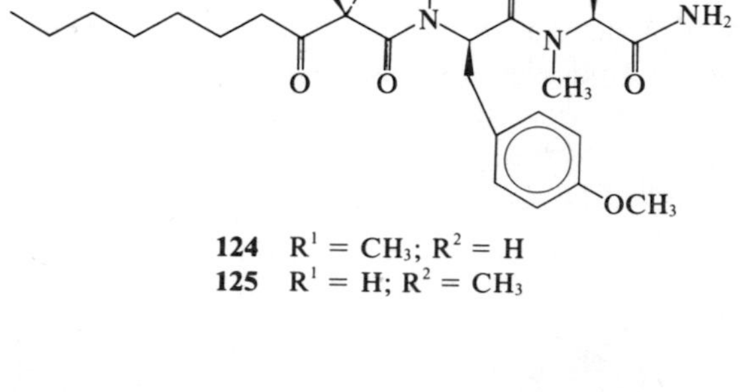

124 $R^1 = CH_3$; $R^2 = H$
125 $R^1 = H$; $R^2 = CH_3$

dimethyltyrosyl unit are present. Acid hydrolysis (1.5 N HCl, 25% ethanol–water, reflux 24 hr) leads to L-alanine, glycine, L-N-methylvaline, L-N,O-dimethyltyrosine (not D as found in **124** and **125**), L-N-methylisoleucine, 2-hydroxy-3-methylpentanoylglycine (**126**), 2-amino-3-oxo-4-methylpentane (**127,** isolated as the hydrochloride) and 3-amino-2-methylpentanoic acid (**128**). The hydrolysis products suggest that majusculamide C is a cyclic nonadepsipeptide composed of subunits

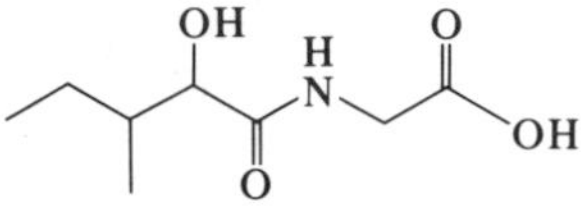

126 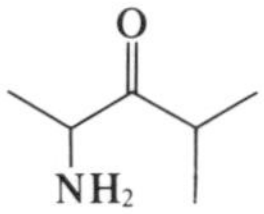   **127** 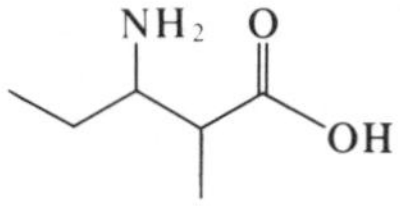   **128**

129a–129h. These seven subunits account for all the atoms in the molecular formula. The mass spectrum indicates that a glycyl-N-methylisoleucyl-glycyl-N-methylvalyl-N,O-dimethyltyrosyl sequence is present.

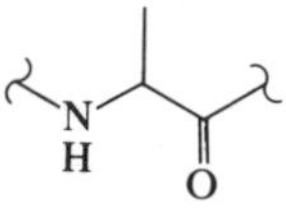

129a 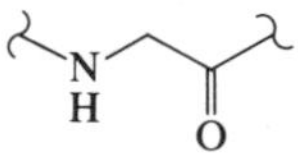   **129b** 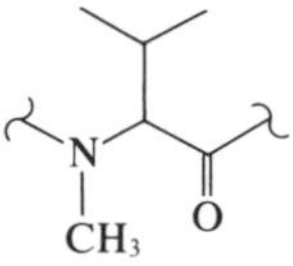   **129c** 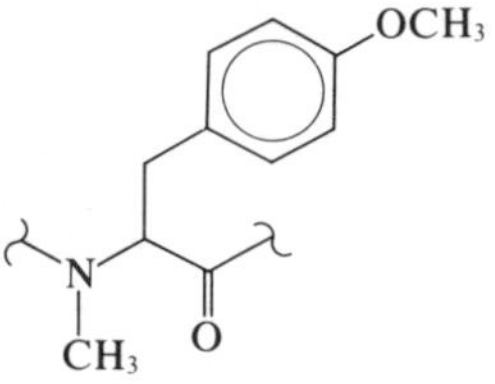   **129d**

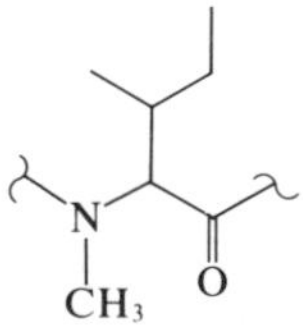

129e 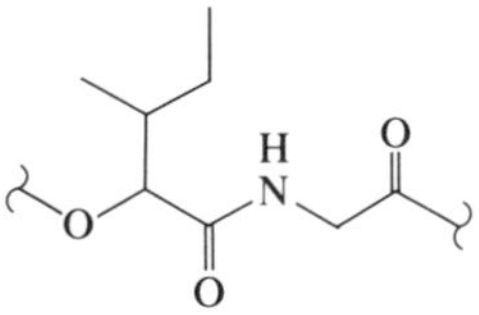   **129f** 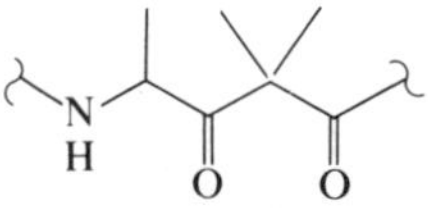   **129g** 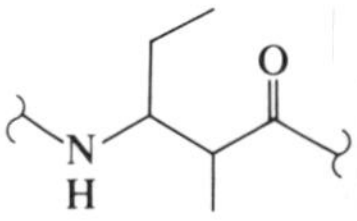   **129h**

D. Miscellaneous Compounds

(+)-$\alpha(S)$-Butyramido- γ-butyrolactone (**130**) is a minor constituent in a shallow-water variety of *L. majuscula* from Kahala Beach, Oahu (Marner and Moore, 1978). The structure was determined from spectral data and confirmed by acid hydrolysis to *n*-butyric acid and α-amino-γ- butyrolactone and synthesis from (−)-$\alpha(S)$-amino-γ-butyrolactone.

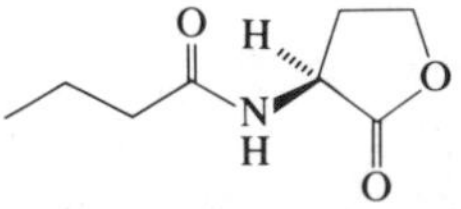

130

Hyellazole (**131**) and 6-chlorohyellazole (**132**) are two unusual nonbasic carbazole alkaloids from a supralittoral variety of the blue-green alga *Hyella caespitosa* (Cardellina *et al.*, 1979a). The structure of 6-chlorohyellazole was solved by X-ray crystallography. The hyellazoles possess structures that are quite different from the carbazole alkaloids of terrestrial plants.

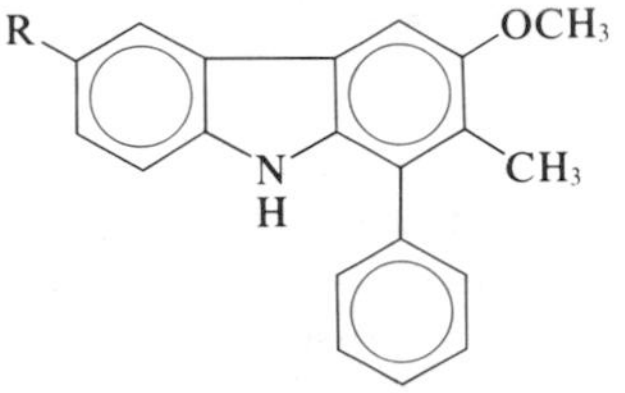

131   R = H
132   R = Cl

ACKNOWLEDGMENTS

The unpublished work done in the author's laboratory on the toxins of blue-green algae was supported by a grant from the National Cancer Institute, DHEW (CA12623-07). All other unpublished work was supported by the National Science Foundation (CHE76-82517). High-frequency nmr studies at the Stanford Magnetic Resonance Laboratory were made possible by NSF grant GP-23633 and NIH grant RR00711.

REFERENCES

Baird, W. M., and Boutwell, R. K. (1971). *Cancer Res.* **31**, 1074.
Banner, A. H. (1959). *Hawaii Med. J.* **19**, 35.
Banner, A. H. (1967). *In* "Animal Toxins" (F. E. Russell and P. R. Saunders, eds.), p. 157. Pergamon, Oxford.
Baptist, J. N., and Werber, F. X. (1964). *SPE Trans.* **4**, 245.
Baslow, M. H. (1977). "Marine Pharmacology." Krieger Publ., Huntington, New York.
Bates, H. A., and Rapoport, H. (1979). *J. Am. Chem. Soc.* **101**, 1259.
Bennett, A., and Bogorad, L. (1971). *Biochemistry* **10**, 3625.
Bird, C. W., Lynch, J. M., Pitt, S. J., Reid, W. W., Brooks, C. J. W., and Middleditch, B. S. (1971a). *Nature (London)* **230**, 473.
Bird, C. W., Lynch, J. M., Pirt, S. J., and Reid, W. W. (1971b). *Tetrahedron Lett.* p. 3189.
Bishop, C. T., Anet, E. F. L. J., and Gorham, P. R. (1959). *Can. J. Biochem. Physiol.* **37**, 453.
Bordner, J., Thiessen, W. E., Bates, H. A., and Rapoport, H. (1975). *J. Am. Chem. Soc.* **97**, 6008.
Boyer, G. L., Schantz, E. J., and Schnoes, H. K. (1978). *Chem. Commun.* p. 889.
Brockman, H. Jr., and Knobloch, G. (1973). *Chem. Ber.* **106**, 803.
Burkholder, P. R., Burkholder, L. M., and Almodovar, L. R. (1960). *Bot. Mar.* **2**, 149.
Byfield, P. G. H., and Zuber, H. (1972). *FEBS Lett.* **28**, 36.

Campbell, H. F., Edwards, O. E., and Kolt, R. (1977). *Can. J. Chem.* **55**, 1372.

Cardellina II, J. H., and Moore, R. E. (1979). *Tetrahedron Lett.* p. 2007.

Cardellina II, J. H., and Moore, R. E. (1980). *Tetrahedron* **36**, 993(1980).

Cardellina II, J. H., Dalietos, D., Marner, F-J., Mynderse, J. S., and Moore, R. E. (1978). *Phytochemistry* **17**, 2091.

Cardellina II, J. H., Kirkup, M. P., Moore, R. E., Mynderse, J. S., Seff, K., and Simmons, C. J. (1979a). *Tetrahedron Lett.* p. 4915.

Cardellina II, J. H., Marner, F-J., and Moore, R. E. (1979b). *J. Am. Chem. Soc.* **101**, 240.

Cardellina II, J. H., Marner, F-J., and Moore, R. E. (1979c). *Science* **204**, 193.

Cardellina II, J. H., Moore, R. E., Arnold, E. V., and Clardy, J. (1979d). *J. Org. Chem.* **44**, 4039.

Carmichael, W. W., and Gorham, P. R. (1978). *Mitt. Int. Verein. Limnol.* **21**, 285.

Carmichael, W. W., Biggs, D. F., and Gorham, P. R. (1975). *Science* **187**, 542.

Carmichael, W. W., Biggs, D. F., and Peterson, M. A. (1979). *Toxicon* **17**, 229.

Carr, N. G. (1966). *Biochim. Biophys. Acta* **120**, 308.

Chapman, D. J., Cole, W. J., and Siegelman, H. W. (1967). *J. Am. Chem. Soc.* **89**, 5976.

Charles, C., Braekman, J. C., Daloze, D., Tursch, B., and Karlsson, R. (1978). *Tetrahedron Lett.* 1519.

Cohn, F. (1871/1872). *Jb. Schles. Ges. Vaterl. Kult.* **49**, 83.

Cole, W. J., Chapman, D. J., and Siegelman, H. W. (1968). *Biochemistry* **7**, 2929.

Cornibert, J., and Marchessault, R. H. (1972). *J. Mol. Biol.* **71**, 735.

Crespi, H. L., and Katz, J. J. (1969). *Phytochemistry* **8**, 759.

Crespi, H. L., Smith, U., and Katz, J. J. (1968). *Biochemistry* **7**, 2232.

Dawes, E. A., and Gibbons, D. W. (1964). *Bacteriol. Rev.* **28**, 126.

De Rosa, M., Gambacorta, A., Minale, L., and Bu'Lock, J. D. (1971). *Chem. Commun.* 619.

Desikachary, T. V. (1973). *In* "The Biology of Blue-Green Algae" (N. G. Carr and B. A. Whitton, eds.), p. 473. Univ. of California Press, Berkeley, California.

De Souza, N. J., and Nes, W. R. (1968). *Science* **162**, 363.

Devlin, J. P., Edwards, O. E., Gorham, P. R., Hunter, N. R., Pike, R. K., and Stavric, B. (1977). *Can. J. Chem.* **55**, 1367.

Drouet, F. (1968). "Revision of the Classification of the Oscillatoriaceae," Monograph 15. Academy of Natural Sciences, Philadelphia, Pennsylvania.

Dunn, D. F., Kashiwagi, M., and Norton, T. R. (1975). *Comp. Biochem. Physiol.* **50**, 133.

Erwin, J., and Bloch, K. (1963). *Biochem. Z.* **338**, 496.

Fogg, G. E., Stewart, W. D. P., Fay, P., and Walsby, A. E. (1973). "The Blue-Green Algae." Academic Press, New York.

Förster, H. J., Biemann, K., Haigh, W. G., Tattrie, N. H., and Colvin, J. R. (1973). *Biochem. J.* **135**, 133.

Fujiki, H., Mori, M., Nakayasu, M., Terada, M., and Sugimura, T. (1979). *Biochem. Biophys. Res. Commun.* **90**, 976.

Gerber, N. N. (1968). *Tetrahedron Lett.* p. 2971.

Glazer, A. N., and Cohen-Bazire, G. (1971). *Proc. Natl. Acad. Sci. U.S.* **68**, 1398.

Glazer, A. N., and Fang, S. (1973). *J. Biol. Chem.* **248**, 659.

Gorham, P. R., and Carmichael, W. W. (1979). *Pure Appl. Chem.* **52**, 165.

Gossauer, A., and Hinze, R-P. (1978). *J. Org. Chem.* **43**, 283.

Gossauer, A., and Weller, J-P. (1978). *J. Am. Chem. Soc.* **100**, 5928.

Herdman, M., Janvier, M., Rippka, R., and Stanier, R. Y. (1979a). *J. Gen. Microbiol.* **111**, 73.

Herdman, M., Janvier, M., Waterbury, J. B., Rippka, R., and Stanier, R. Y. (1979b). *J. Gen. Microbiol.* **111**, 63.

Hofheinz, W., and Oberhänsli, W. E. (1977). *Helv. Chim. Acta* **60**, 660.

Holton, R. W., Blecker, H. H., and Stevens, T. S. (1968). *Science* **160**, 545.

Huber, C. S. (1972). *Acta Crystallogr.* **B28**, 2577.

Jenkins, D., Medsker, L. L., and Thomas, J. F. (1967). *Environ. Sci. Technol.* **1**, 731.

Jensen, T. E., and Sicko, L. M. (1971). *J. Bact.* **106**, 683.

Jüttner, F. (1976). *Z. Naturforsch.* **31c**, 491.

Kashiwagi, M., Mynderse, J. S., Moore, R. E., and Norton, T. R. (1980). *J. Pharm. Sci.* **69**, 735.

Kato, Y. (1973). PhD Dissertation, Univ. of Hawaii, Honolulu.

Kato, Y., and Scheuer, P. J. (1975). *Pure Appl. Chem.* **41**, 1.

Kato, Y., and Scheuer, P. J. (1976). *Pure Appl. Chem.* **48**, 29.

Kazlauskas, R., Lidgard, R. O., Wells, R. J., and Vetter, W. (1977). *Tetrahedron Lett.* p. 3183.

Kazlauskas, R., Murphy, P. T., Wells, R. J., Daly, J. J., and Schönholzer, P. (1978a). *Tetrahedron Lett.* 4951.

Kazlauskas, R., Murphy, P. T., and Wells, R. J. (1978b). *Tetrahedron Lett.* p. 4945.

Keleti, G., Sykora, J. L., Lippy, E. C., and Shapiro, M. A. (1979). *Appl. Environ. Microbiol.* **38**, 471.

Kirpenko, Yu, A., Perevozchenko, I. I., Sirenko, K. A., and Lukina, L. F. (1975). *Dopov. Akad. Nauk. Ukr. RSR Ser. B* 359.

Köst-Reyes, E., Köst, H-P., and Rüdiger, W. (1975). *Justus Liebigs Ann. Chem.* p. 1594.

Lagarias, J. C., Glazer, A. N., and Rapoport, H. (1979). *J. Am. Chem. Soc.* **101**, 5030.

Lemoigne, M. (1925). *Ann. Inst. Pasteur Paris* **39**, 144.

Lemoigne, M. (1927). *Ann. Inst. Pasteur Paris* **41**, 148.

Levin, E. Y., and Bloch, K. (1964). *Nature (London)* **202**, 90.

Liaaen-Jensen, S. (1978). *In* "Marine Natural Products: Chemical and Biological Perspectives" (P. J. Scheuer, ed.), Vol. 2, p. 1. Academic Press, New York.

Libby, E. C., and Erb, J. (1976). *J. Am. Water Works Assoc.* **68**, 606.

Lightner, D. V. (1978). *J. Invertebrate Pathol.* **32**, 139.

Loui, M. S. M., and Moore, R. E. (1979). *Phytochemistry* **18**, 1733.

Marchessault, R. H., Okamura, K., and Su, C. J. (1970). *Macromolecules* **3**, 735.

Marner, F-J., and Moore, R. E. (1978). *Phytochemistry* **17**, 553.

Marner, F-J., Moore, R. E., Hirotsu, K., and Clardy, J. (1977). *J. Org. Chem.* **42**, 2815.

Medsker, L. L., Jenkins, D., and Thomas, J. F. (1968). *Environ. Sci. Technol.* **2**, 461.

Merrick, J. M., and Doudoroff, J. (1964). *J. Bacteriol.* **88**, 60.

Moore, R. E. (1977). *Bioscience* **27**, 797.

Moore, R. E. (1978). *In* "Marine Natural Products: Chemical and Biological Perspectives" (P. J. Scheuer, ed.), Vol. I, p. 43. Academic Press, New York.

Mynderse, J. S., and Moore, R. E. (1978a). *J. Org. Chem.* **43**, 2301.

Mynderse, J. S., and Moore, R. E. (1978b). *J. Org. Chem.* **43**, 4359.

Mynderse, J. S., and Moore, R. E. (1978c). *Phytochemistry* **17**, 1325.

Mynderse, J. S., and Moore, R. E. (1979). *Phytochemistry* **18**, 1181.

Mynderse, J. S., Moore, R. E., Kashiwagi, M., and Norton, T. R. (1977). *Science* **196**, 538.

Nakata, H., Harada, H., and Hirata, Y. (1966). *Tetrahedron Lett.* p. 2515.

Nichols, B. W. (1973). *In* "The Biology of Blue-Green Algae" (N. G. Carr and B. A. Whitton, eds.), p. 144. Univ. of California Press, Berkeley, California.

O'Carra, P., and Killilea, S. D. (1971). *Biochem. Biophys. Res. Commun.* **45**, 1192.

RamaMurthy, J., and Capindale, J. B. (1970). *Can. J. Biochem.* **48**, 508.

Reitz, R. C., and Hamilton, J. G. (1968). *Comp. Biochem. Physiol.* **25**, 401.

Rippka, R., Dervelles, J., Waterbury, J. B., Herdman, M., and Stanier, R. Y. (1979). *J. Gen. Microbiol.* **111**, 1.

Rohmer, M., and Ourisson, G. (1976). *Tetrahedron Lett.* pp. 3633, 3637.

Rose, A. F., Scheuer, P. J., Springer, J. P., and Clardy, J. (1978). *J. Am. Chem. Soc.* **100**, 7665.

Rüdiger, W., and O'Carra, P. (1969). *Eur. J. Biochem.* **7**, 509.

Rüdiger, W., O'Carra, P., and O'hEocha, C. (1967). *Nature (London)* **215**, 1477.

Runnegar, M. T. C., and Falconer, I. R. (1975). *Proc. Austral. Biochem. Soc.* **8**, 5.

Russell, D. H., and Levy, C. C. (1971). *Cancer Res.* **31**, 248.

Safferman, R. S., Rosen, A. A., Mashni, C. I., and Morris, M. E. (1967). *Environ. Sci. Technol.* **1**, 429.

Sakabe, N., Harada, H., Hirata, Y., Tomiie, Y., and Nitta, I. (1966). *Tetrahedron Lett.* p. 2523.

Schantz, E. J., Ghazarossian, V. E., Schnoes, H. K., Strong, F. M., Springer, J. P., Pezzanite, J. O., and Clardy, J. (1975). *J. Am. Chem. Soc.* **97**, 1238.

Schubert, K., Rose, G., Wachtel, H., Horhold, C., and Ikekawa, N. (1968). *Eur. J. Biochem.* **5**, 246.

Schwimmer, D., and Schwimmer, M. (1964). *In* "Algae and Man (D. F. Jackson, ed.), p. 368. Plenum Press, New York.

Sharma, G. M., and Vig, B. (1972). *Tetrahedron Lett.* 1715.

Sharma, G. M., Vig, B., and Burkholder, P. R. (1969). *Food-Drugs Sea Conf., 2nd, 1968* p. 307.

Shelton, J. R., Agostini, D. E., and Lando, J. B. (1971a). *J. Polym. Sci. Chem. Ed.* **9**, 2789.

Shelton, J. R., Lando, J. B., and Agostini, D. E. (1971b). *Polym. Lett.* **9**, 173.

Shimizu, Y. (1978). *In* "Marine Natural Products: Chemical and Biological Perspectives" (P. J. Scheuer, ed.), Vol. 1, p. 1. Academic Press, New York.

Shimizu, Y., Buckley, L. J., Alam, M., Oshima, Y., Fallon, W. E., Kasai, H., Miura, I., Gullo, V. P., and Nakanishi, K. (1976). *J. Am. Chem. Soc.* **98**, 5414.

Shimizu, Y., Hsu, C-P., Fallon, W. E., Oshima, Y., Miura, I., and Nakanishi, K. (1978). *J. Am. Chem. Soc.* **100**, 6791.

Simmons, C. J., Marner, F-J., Cardellina II, J. H., Moore, R. E., and Seff, K. (1979). *Tetrahedron Lett.* p. 2003.

Smith, G. M. (1938). "Cryptogamic Botany," Vol. I. McGraw-Hill, New York.

Solomon, A. E., and Stoughton, R. B. (1978). *Arch. Dermatol.* **114**, 1333.

Stanier, R. Y., and van Niel, C. B. (1962). *Arch. Mikrobiol.* **42**, 17.

Starr, T. J., Dieg, E. F., Church, K. K., Allen, M. B. (1962). *Tex. Rep. Biol. Med.* **20**, 271.

Stewart, W. D. P. (1973). *In* "The Biology of Blue-Green Algae" (N. G. Carr and B. A. Whitton, eds.), p. 260. Univ. of California Press, Berkeley, California.

Sverdrup, H. V., Johnson, M. W., and Fleming, R. H. (1947). "The Oceans." Prentice-Hall, Englewood Cliffs, New Jersey.

Tabrah, F. L., Kashiwagi, M., and Norton, T. R. (1972). *Int. J. Clin. Pharmacol.* **5**, 420.

Takashima, M., Sakai, H., Mori, R., and Arima, K. (1962). *Agr. Biol. Chem.* **26**, 669.

Tanino, H., Nakata, T., Kaneko, T., and Kishi, Y. (1977). *J. Am. Chem. Soc.* **99**, 2818.

Troxler, R. F., Brown, A., Lester, R., and White, P. (1970). *Science* **167**, 192.

Troxler, R. F., and Dokos, J. M. (1973). *Plant Physiol.* **51**, 72.

Vernon, L. P., and Seely, G. R. (1966). "The Chlorophylls." Academic Press, New York.

Watson, M. (1973). *Toxicon* **11**, 259.

Watson, M., and Rayner, M. D. (1973). *Toxicon* **11**, 269.

Williams, V. P., and Glazer, A. N. (1978). *J. Biol. Chem.* **253**, 202.

Williamson, D. H., and Wilkinson, J. F. (1958). *J. Gen. Microbiol.* **19**, 198.

Chapter 2

Guanidine Derivatives

L. CHEVOLOT

Nitrogen-containing marine compounds have received relatively less attention than terpenes, sterols, and other lipid substances. Nevertheless, a wide variety of nitrogen-containing products have been isolated from marine organisms, many of which are not found in the terrestrial flora or fauna. Among the nitrogen-containing compounds, those which contain a guanidine unit are particularly interesting and are often unique to marine organisms, as pointed out by Baker and Murphy (1976). In addition, a

MARINE NATURAL PRODUCTS

ISBN 0-12-624004-3

large number of these compounds possess powerful biological activities. The present chapter is confined to a discussion of these guanidine derivatives. Basic peptides containing arginine are not considered here.

I. SIMPLE LINEAR GUANIDINE DERIVATIVES

Many simple products of the linear guanidine class are widely distributed in both terrestrial and marine environments. However, many interesting compounds of a more complex nature have been isolated only from marine sources. In fact, the greatest product diversity is observed in the marine worms of the phyla Annelida, Sipunculida, and Echiura. The guanidino compounds from terrestrial worms also display several particularly interesting features. Some well-documented reviews have previously been published in this field (Thoai, 1965; Thoai and Robin, 1969; Needham, 1970). For such widely distributed compounds as arginine (**1**), agmatine (**2**), and creatine (**3**) we refer the reader to these publications and to the compilation of marine natural products published by Baker and Murphy (1976). It should be mentioned that many of the compounds described in this section were discovered by researchers of the Collège de France (Roche, Thoai, Robin, and co-workers); few other investigators have shown an interest in these compounds.

$$H_2N-\underset{\underset{NH}{\|}}{C}-NH-(CH_2)_3-CH(NH_2)-COOH$$

1

$$H_2N-\underset{\underset{NH}{\|}}{C}-NH-(CH_2)_4NH_2$$

2

$$H_2N-\underset{\underset{NH}{\|}}{C}-NMe-CH_2-COOH$$

3

A. Simple Arginine Derivatives

Arginine, of course, yields a great diversity of oxidation or degradation products. Hydroxyarginine derivatives are often found in the seeds of higher plants (Bell, 1964; Bell and Tirimana, 1964); however, such compounds have not often been reported in marine organisms. Makisumi

(1961) found γ-hydroxyarginine (**4**), hydroxyagmatine (**5**), and β-hydroxy-γ-guanidinobutyric acid (**6**) in addition to nonhydroxylated compounds in the sea anemone *Anthopleura japonica*. Fujita (1959, 1960, 1961) isolated γ-hydroxyarginine (**4**) from the sea cucumber *Polycheira rufescens*. However, these products have not been actively researched and are perhaps more widely distributed in the marine environment than

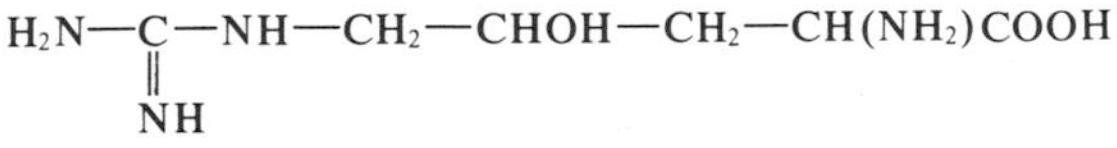

$$H_2N—C—NH—CH_2—CHOH—CH_2—CH(NH_2)COOH$$
$$\|$$
$$NH$$

4

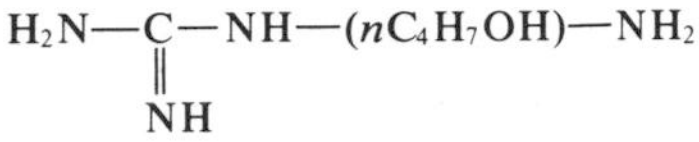

$$H_2N—C—NH—(nC_4H_7OH)—NH_2$$
$$\|$$
$$NH$$

5

$$H_2N—C—NH—CH_2—CHOH—CH_2—COOH$$
$$\|$$
$$NH$$

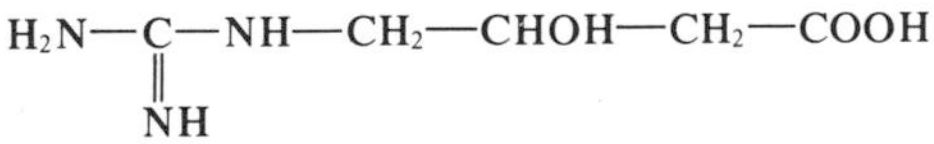

6

these reports would indicate. α-Keto-δ-guanidinovaleric acid (**7**) is an oxidation product of arginine often found in echinoderms, mollusks, and crustaceans (Thoai *et al.*, 1953a). However, its presence was not detected in marine worms. Its oxidative decarboxylation product, γ-guanidinobutyric acid (**8**) is encountered in many marine invertebrates (Thoai *et al.*, 1953a) and also in the seaweed *Gymnogongrus flabelliformis* (Ito *et al.*, 1967). Agmatine (**2**) is a relatively common compound (Baker

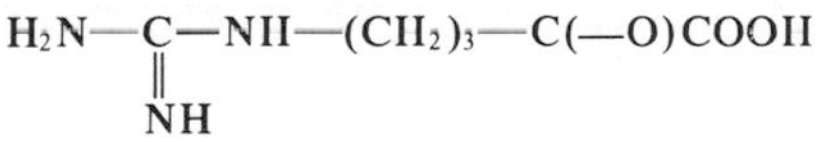

$$H_2N—C—NH—(CH_2)_3—C(—O)COOH$$
$$\|$$
$$NH$$

7

$$H_2N—C—NH—(CH_2)_3—COOH$$
$$\|$$
$$NH$$

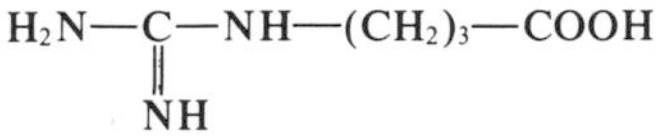

8

and Murphy, 1976), but the presence of *N*-methylagmatine (**9**) has seldom been noted. Florkin and Bricteux-Gregoire (1972, p. 313) cited only one reference (Iseki, 1931), which reported the presence of this compound in octopus muscle. Haurowitz and Waelsch (1926) found a guanidino compound which may be *N*-methylagmatine (**9**), in the medusoid *Velella spirans*. *N*-Acetylagmatine (**10**) has recently been isolated from the sea anemones *Actinia equina* and *A. fragacea* (Guillou and Robin, 1979).

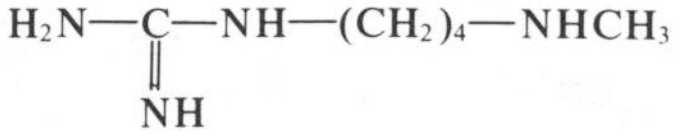

$$H_2N—\underset{\underset{NH}{\|}}{C}—NH—(CH_2)_4—NHCH_3$$

9

$$H_2N—\underset{\underset{NH}{\|}}{C}—NH—(CH_2)_1—NHCOCH_3$$

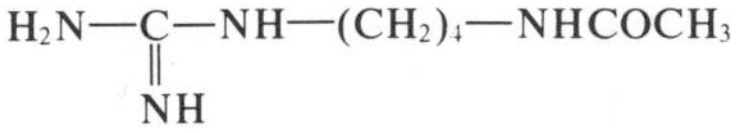

10

B. Sulfur-Containing Guanidino Compounds

All compounds of this class are derived from the parent compound β-aminoethanesulfonic acid (taurine). The first compound isolated in this series was asterubin (**11**). Ackermann (1935) obtained this product from the sea stars *Marthasterias glacialis* and *Asterias rubens*. Asterubin is one of the rare noncyclic guanidino compounds substituted on two nitrogens. It was synthesized by Ackermann and Müller (1935), but no further work has been done on this product, although its concentration in sea stars is relatively high (about 0.01% of wet weight).

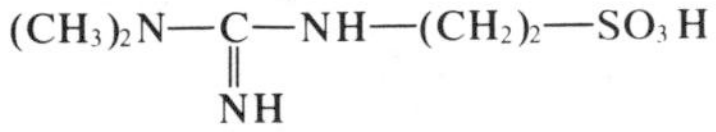

$$(CH_3)_2N—\underset{\underset{NH}{\|}}{C}—NH—(CH_2)_2—SO_3H$$

11

Taurocyamine (**12**) was discovered in the sedentary polychaete *Arenicola marina* (Thoai *et al.*, 1953b; Thoai and Robin, 1954a). In fact, this compound is widely distributed among invertebrates, e.g., worms (Thoai and Robin, 1969), sponges (Roche and Robin, 1954; Ackermann and Pant, 1961; Bergquist and Hartmann, 1969), and sea anemones (Makisumi, 1961). It also occurs in small quantities in vertebrates. This compound is a phosphagen precursor (see Section II); however, its presence is not necessarily accompanied by the corresponding phosphagen.

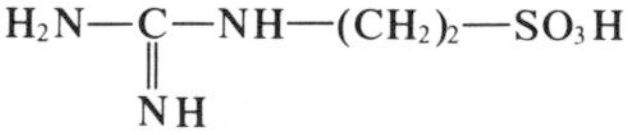

$$H_2N—\underset{\underset{NH}{\|}}{C}—NH—(CH_2)_2—SO_3H$$

12

Hypotaurocyamine (**13**), the more highly reduced derivative, is encountered in annelids and in sipunculids (Robin and Thoai, 1962; Thoai and Robin, 1969). In sipunculids, **13** is found in relatively high concentration.

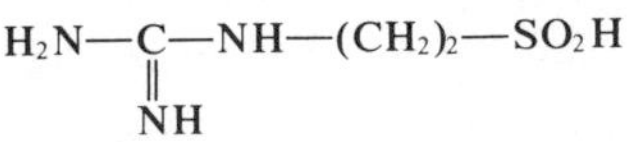

$$H_2N—\underset{\underset{NH}{\|}}{C}—NH—(CH_2)_2—SO_2H$$

13

In the latter taxa, this usually constitutes the main guanidine compound, accompanied by its phosphorylated derivative. There is an exception in this phylum: *Sipunculus nudus* contains arginine (1) as phosphagen base rather than hypotaurocyamine. In annelids and sipunculids, hypotaurocyamine is biosynthesized through transamidination between arginine and the corresponding amine (Thoai *et al.*, 1963a). Taurocyamine (12), on the other hand, is synthesized by oxidation of the hypotaurocyamine sulfinic to the sulfonic function.

C. Linear Diguanidino Derivatives

Three linear diguanidino compounds are presently known, two of which are found in marine organisms. The first isolation of arcaine (14) was made from the muscle of the mollusk *Arca noae* (Kutscher *et al.*, 1931). Arcaine was subsequently found in the mollusk *Cristaria plicata* (Suzuki and Muraoka, 1954) and in various marine worms such as *Audouinia tentaculata* (Robin *et al.*, 1956), as well as in various freshwater worms such as *Hirudo medicinalis* (Robin *et al.*, 1957; Robin and Roche, 1965). As

$$H_2N-\underset{\underset{NH}{\|}}{C}-NH-(CH_2)_4-NH-C(=NH)-NH_2$$

14

yet, the homologous compound audouine (15) has only been isolated in the sedentary marine polychaete *A. tentaculata* (Roche *et al.*, 1965), whereas the parent product hirudonine (16) or diamidino spermidine, is only found in freshwater and terrestrial worms (Robin and Thoai, 1961a; Robin and Roche, 1965).

$$H_2N-C(=NH)-NH-(CH_2)_5-NH-C(=NH)-NH_2$$
15

$$H_2N-C(=NH)-NH-(CH_2)_3-NH-(CH_2)_4-NH-C(=NH)NH_2$$
16

The biogenesis of these compounds has been well studied by the Collège de France group. These researchers showed that arcaine (14), audouine (15) and hirudonine (16) are biosynthesized by the transfer of two amidine groups from arginine to putrescine, cadaverine, and spermidine, respectively (Thoai and Robin, 1969, p. 181).

D. Gongrine and Gigartinine

In the mid-1960s Ito and Hashimoto (1965, 1966 a,b) isolated two unique metabolites, gongrine (17) and gigartinine (18), from the red alga

$$H_2N—C—NH—CO—NH—(CH_2)_3—COOH$$
$$\quad\quad\|$$
$$\quad\quad NH$$

17

$$H_2N—C—NH—CO—NH—(CH_2)_3—CH—COOH$$
$$\quad\quad\|\quad\quad\quad\quad\quad\quad\quad\quad\quad |$$
$$\quad\quad NH\quad\quad\quad\quad\quad\quad\quad\quad NH_2$$

18

Gymnogongrus flabelliformis. Their structures were established mainly by chemical means and confirmed by synthesis of both compounds by known procedures (Ito and Hashimoto, 1969). Ito *et al.* (1966) also searched for these compounds in 25 species of brown, green, and red algae. With the exception of one dubious case, none of the green or brown algae were found to contain these new compounds. On the other hand, four species of red algae contain both products and four contain only gigartinine (**18**). This raises the possibility that gongrine (**17**) is derived from gigartinine (**18**), just as γ-guanidinobutyric acid (**8**) is derived from arginine (**1**).

E. Octopine

Octopine (**19**) is yet another derivative of arginine. It was first isolated by Morizawa (1927) from octopus muscle and subsequently from many species of mollusks (Robin and Guillou, 1977). It has also been found in one species of sipunculid (*Sipunculus nudus;* Thoai and Robin, 1959a), and in two species of Californian nemertean worms (*Cerebratus occidentalis* and *Lineus pictifrons;* Robin, 1964a).

$$H_2N—C—NH—(CH_2)_3—CH—NH—CHMe—COOH$$
$$\quad\quad\|\quad\quad\quad\quad\quad\quad\quad |$$
$$\quad\quad NH\quad\quad\quad\quad\quad\quad COOH$$

19

It has been suggested that the presence of high concentrations of octopine in the muscles of some species signifies that octopine is a precursor of a new muscular phosphagen (Moore and Wilson, 1937); however, attempts to isolate the phosphagen have been unsuccessful. In a series of experiments Thoai and Robin demonstrated that octopine did play a role during muscular contraction, although this role was different from the one previously proposed. Initially, these authors (Thoai and Robin, 1959a) confirmed the previously proposed scheme of the biosynthesis of octopine (**19**) (see Fig. 1). They further showed that octopine synthesis requires an enzyme with NADH as cofactor; they therefore hypothesized that octopine plays an important role during anaerobic

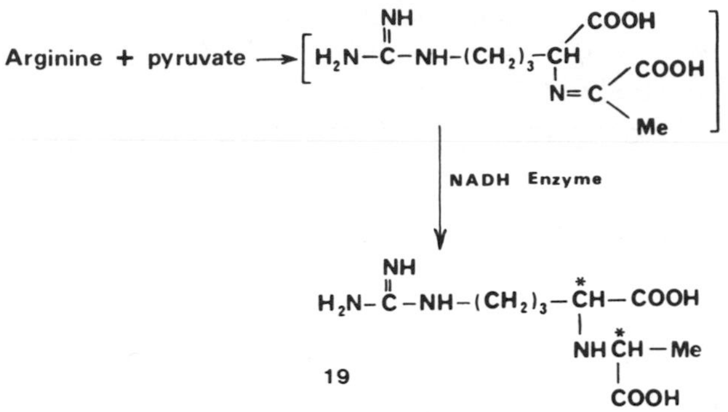

$$\text{Arginine} + \text{pyruvate} \longrightarrow \left[\underset{\overset{\|}{\text{H}_2\text{N}-\text{C}-\text{NH}-(\text{CH}_2)_3-\underset{\underset{\text{N}=\text{C}\underset{\backslash}{\overset{/\text{COOH}}{}}}{|}}{\text{CH}}\overset{/\text{COOH}}{}}}{\overset{\text{NH}}{}} \right]$$

NADH   Enzyme

$$\underset{19}{\overset{\text{NH}}{\underset{\overset{\|}{}}{\text{H}_2\text{N}-\text{C}-\text{NH}-(\text{CH}_2)_3-\overset{*}{\text{CH}}-\text{COOH}}}}$$
$$\underset{\overset{|}{\text{COOH}}}{\text{NH}\,\overset{*}{\text{CH}}-\text{Me}}$$

Fig. 1. Biosynthesis of octopine.

glycolysis through the elimination of pyruvate. In subsequent work these authors demonstrated that the concentration of octopine is remarkably constant in fresh tissues; they postulated, therefore, the presence of a regulatory system (Thoai and Robin, 1959b). In fact, the synthesis of octopine is reversible and requires a new enzyme, octopine dehydrogenase (Thoai and Robin, 1961). This enzyme was purified and its catalytic properties further studied (Thoai *et al.*, 1969; Doublet *et al.*, 1975; Gäde and Grieshaber, 1975).

In a supplementary paper, Robin and Thoai (1961b) demonstrated that those species which contain octopine have very low lactate dehydrogenase activity. These results confirmed the dual role of octopine during muscular contraction: (1) reversible storage of arginine, which results from the degradation of phosphoarginine, and (2) the utilization of pyruvate during anaerobic glycolysis and regeneration of NAD$^+$ without production of lactate.

Support for this proposed role of octopine comes from a recent series of papers (Gäde *et al.*, 1978; Baldwin and Opie, 1978 and references cited therein) in which the relationship between phosphagen depletion, octopine formation, and NAD$^+$ regeneration in muscle and nonmuscle tissues from several mollusks has been further investigated. Regeneration of NAD$^+$ during anoxic muscular work seems to be the more important function of octopine dehydrogenase. It would be interesting to know if similar compounds such as strombine (CH_3—$CH(COOH)$—NH—CH_2—COOH) or α-iminodipropionic acid (CH_3—$CH(COOH)$—NH—CH$(COOH)CH_3$) play similar roles to those of octopine. It is intriguing to note that the alanine moiety of the three products (strombine, octopine, α-iminodipropionic acid) possesses the same stereochemical $R(\text{D})$ configuration (Sato *et al.*, 1978). In passing, it is not without interest to report that strombine (first isolated from the conch *Strombus gigas*)

induces exploratory feeding behavior in some fishes at 10^{-8} g/liter
(Sangster *et al.*, 1975). The α-iminodipropionic acid has been isolated
from muscle extracts of the squid *Todares pacificus* (Sato *et al.*, 1977); the
properties of this compound remain unknown. The biogenesis of strombine, an α-iminodipropionic acid, could be similar to that of octopine, but
no study has been performed.

F. Lombricine, Opheline, Thalessemine, and Bonellidine

In contrast to octopine (**19**), lombricine (**20**), opheline (**21**), and thalassemine (**22**) are phosphagen precursors (see Section II). These three

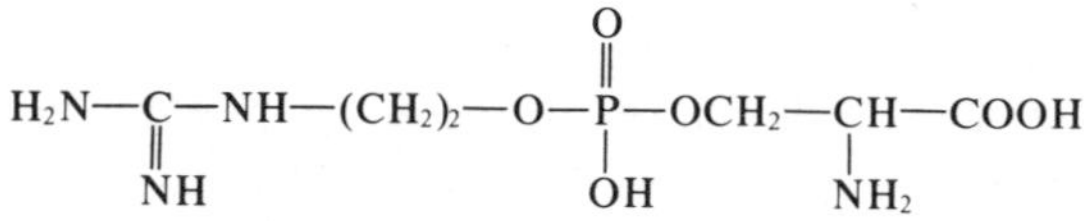

$$H_2N-\underset{\underset{NH}{\|}}{C}-NH-(CH_2)_2-O-\underset{\underset{OH}{|}}{\overset{\overset{O}{\|}}{P}}-OCH_2-\underset{\underset{NH_2}{|}}{CH}-COOH$$

20

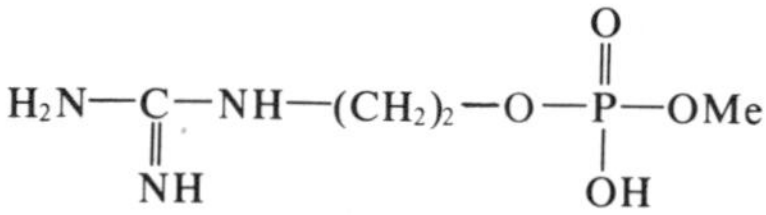

$$H_2N-\underset{\underset{NH}{\|}}{C}-NH-(CH_2)_2-O-\underset{\underset{OH}{|}}{\overset{\overset{O}{\|}}{P}}-OMe$$

21

$$H_2N-\underset{\underset{NH}{\|}}{C}-NH-(CH_2)_2-O-\underset{\underset{OH}{|}}{\overset{\overset{O}{\|}}{P}}-OCH_2-\underset{\underset{NMe_2}{|}}{CH}-COOH$$

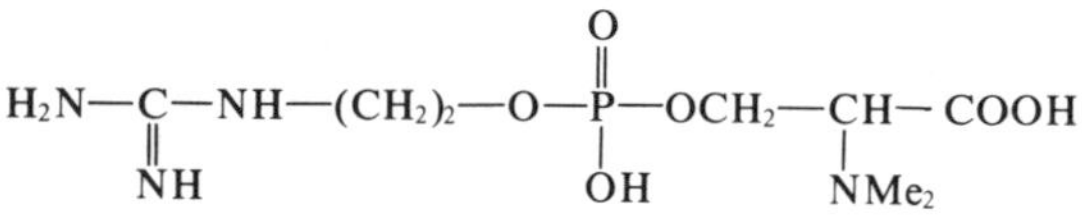

22

compounds are closely related. They are derived from the same
guanidinoethyl phosphate residue and differ in substitution on the phosphate group: methyl in opheline, seryl in lombricine, (α-*N,N*-di-
methyl)seryl in thalassemine. Of these compounds, lombricine (**20**) was
first isolated from the earthworm *Lumbricus terrestris* (Thoai and Robin,
1954b), and its structure was proposed largely on the basis of identification of hydrolysis products. The structure was later confirmed by synthesis (Beatty and Magrath, 1959). These workers in addition established the
D configuration of the serine residue (Beatty *et al.*, 1959). In fact, lombricine (**20**) is present in many terrestrial and marine worms, especially in
oligochaetes, but also in polychaetes and in two echiurids (Thoai and
Robin, 1969). On the other hand, opheline (**21**) and thalassemine (**22**) are
found only in marine worms. Opheline is found in the sedentary
polychaete *Ophelia neglecta* (Thoai *et al.*, 1963b), while thalassemine is

present in the echiurid *Thalassema neptuni* (Thoai *et al.*, 1972). In both cases, synthesis confirmed the proposed structures. In this series of compounds (lombricine and thalassemine) the configuration of the serine residue depends on the organism from which the compound originates. Serine exists in the D form in Annelida, but in the L form in Echiura (Thoai *et al.*, 1972).

Bonellidine* (**23**) is a more complex compound of this group. It was isolated from an echiurid, *Bonellia viridis,* by Thoai *et al.* (1967). As a result of a study of the hydrolysis products, two formulas, **23** and **24,** have been proposed. The authors prefer structure **23** on the basis of color reactions. The stereochemistry of the serine residue has been determined by enzymatic methods (Robin and Guillou, 1980) to be L. This agrees with the rule deduced by Thoai *et al.* (1972).

$$HO_2C—CH(NH_2)—CH_2—CONH—CH(CO_2H)—CH_2—OP(=O)(OH)—O(CH_2)_2—NH—C\underset{NH_2}{\overset{NH}{\lessgtr}}$$

23

$$HO_2C—CH_2—CH(NH_2)CONH—CH(CO_2H)—CH_2—OP(=O)(OH)—O(CH_2)_2—NH—C\underset{NH_2}{\overset{NH}{\lessgtr}}$$

24

Bonellidine is the only guanidino compound present in the trunk muscle of *B. viridis.* Lombricine and arginine are found in other tissues. This observation raises the possibility that bonellidine is a precursor of a new phosphagen, despite the fact that no phosphobonellidine has ever been isolated.

G. Phascoline and Phascolosomine

These compounds are found in very high concentration in the viscera of three species of Sipunculida. *Phascolion strombi* contains phascoline (**25**), while *Phascolosoma vulgare* and *P. elongata* produce phascolosomine (**26**) (Guillou and Robin, 1973). The presence of large quantities of guanidino compounds in the viscera is notable, as the concentration of such products is usually high only in muscles. In both cases, acid hydrolysis gave a guanidino compound (**27** or **28**) and an amine (**29** or **30**). Each product was independently characterized. The proposed structures were confirmed by nmr and mass spectroscopy.

---

*Not to be confused with bonellin, the green pigment isolated by Lederer (1939) and structurally elucidated by Pelter (1976).

$$HN\!\!=\!\!\underset{NH_2}{\overset{}{C}}\!\!-\!\!NH\!\!-\!\!(CH_2)_2\!\!-\!\!CONH\!\!-\!\!CH_2\!\!-\!\!CHOH\!\!-\!\!(CH_2)_4\!\!-\!\!CH_3$$

25

$$\downarrow HCl$$

$$HN\!\!=\!\!\underset{NH_2}{\overset{}{C}}\!\!-\!\!NH\!\!-\!\!(CH_2)_2COOH + H_2N\!\!-\!\!CH_2\!\!-\!\!CHOH\!\!-\!\!(CH_2)_4\!\!-\!\!CH_3$$

27   **29**

$$HN\!\!=\!\!\underset{NH_2}{\overset{}{C}}\!\!-\!\!NH\!\!-\!\!CH_2\!\!-\!\!CHMe\!\!-\!\!CONH\!\!-\!\!CH_2\!\!-\!\!CHOMe\!\!-\!\!(CH_2)_4\!\!-\!\!CH_3$$

26

$$\downarrow HCl$$

$$HN\!\!=\!\!\underset{NH_2}{\overset{}{C}}\!\!-\!\!NH\!\!-\!\!CH_2\!\!-\!\!CHMe\!\!-\!\!COOH + H_2N\!\!-\!\!CH_2\!\!-\!\!CHOMe\!\!-\!\!(CH_2)_4\!\!-\!\!CH_3$$

28   **30**

In fact, β-guanidinoisobutyric acid (**28**) has previously been reported in free form together with the bound form in the viscera of various sipunculids (*Phascolosoma vulgare, P. agassizii, P. elongatum, Dendrostromum dyscritum,* and *Siphonosoma ingens*) and in the tractus of the annelid *Arenicola marina* (Robin, 1964b). Robin (1964b) raised the possibility that β-guanidinoisobutyric acid (**28**) plays a role in the metabolic pathway of pyrimidines. Recently this point was established by *in vivo* studies. The carbon chains of the phascoline and phascolosomine guanidine moieties are products of pyrimidine catabolism; in addition, the amidine group is derived from arginine by transamidination (Robin and Guillou, personal communication).

Baslow (1977) reported that in the mid-1950s Chaet had detected the presence of two toxins in the coelemic fluid (or in the coelemic cells) of *Golfingia* (*Phascolosoma*) *gouldi*. There may be some connection between these toxins and the compounds phascoline and phascolosomine, as both compounds display similar and intriguing effects on cultured rat heart cells (Auclair *et al.*, 1976). At concentrations of 10^{-3} M or higher both compounds stop cell pulsation; this effect is obtained after 30 min for phascolosomine, but phascoline requires 24 hours for complete effect. At concentrations of 10^{-4} M both compounds, after 24 hours of contact, only retard the beat rhythm. These effects are reversible at concentrations of 10^{-3} M or lower, but are irreversible at 10^{-2} M. Similar effects are displayed by the amino alcohol **29** or the amino ether **30** obtained by hydrolysis. The authors thus deduce that activity is due to these residues.

In addition, phascoline and phascolosomine are moderate inhibitors of acetylcholinesterase activity (Matsumoto *et al.*, 1977).

H. Biosynthesis

Briefly, it will be recalled that there are two main biosynthetic pathways for the biological elaboration of these compounds: (1) transfer of an amidine group from arginine to a convenient primary amine (this reaction may be followed by subsequent modifications such as methylation or oxidation) and (2) transformation of arginine itself (oxidation, decarboxylation, degradation of the carbon chain, condensation, etc.). For more information the reader is referred to Thoai and Robin (1969) and Needham (1970). It should be noted that biosynthetic pathways may be different in different phyla. For instance, agmatine (2) is usually biosynthesized by decarboxylation of arginine (1). However, in the leech *Hirudo medicinalis,* this compound is elaborated by amidination of putrescine.

II. PHOSPHAGENS

Phosphagens are an important class of compounds which are present in particularly high concentration in muscles where they serve as an energy reserve for muscular contraction. Some interesting reviews of phosphagens have been published (Thoai and Roche, 1964; Thoai and Robin, 1969; Robin, 1974, 1980). At the present time, eight phosphagens are known: phosphoarginine (31), phosphocreatine (32), phosphoglycocyamine (33), phosphotaurocyamine (34), phosphohypotaurocyamine (35), phosphoopheline (36), phospholombricine (37) and phosphothalassemine (38). The formulas are shown in Table 1. Once again, most of the knowledge in this field is due to the work of the Collège de France group. Six of the eight known phosphagens have been discovered by these researchers.

The most important feature of these compounds is the presence of an energy-rich bond between a guanidyl group and a phosphoryl residue. The lability of this energy-rich bond is due to the large amount of resonance stability of each ion (phosphate and guanidinium) following bond cleavage. As result, phosphagens are capable of phorphorylating ADP to ATP by the reversible reaction:

$$\text{Phosphagen} + \text{ADP} \rightleftharpoons \text{guanidine derivative} + \text{ATP}$$

The level of ATP may thus be maintained during intense muscular work because the other ATP-producing reactions (glycolysis, oxidative phosphorylation) proceed too slowly to furnish the amount of ATP required

TABLE 1

Marine Phosphagens

Compound	Structure
31 Phosphoarginine	$H_2O_3PNH-\overset{\overset{\displaystyle NH}{\|\|}}{C}-NH-(CH_2)_3-CH(NH_2)COOH$
32 Phosphocreatine	$H_2O_3PNH-\overset{\overset{\displaystyle NH}{\|\|}}{C}-NMe-CH_2-COOH$
33 Phosphoglycocyamine	$H_2O_3PNH-\overset{\overset{\displaystyle NH}{\|\|}}{C}-NH-CH_2-COOH$
34 Phosphotaurocyamine	$H_2O_3PNH-\overset{\overset{\displaystyle NH}{\|\|}}{C}-NH-(CH_2)_2-SO_3H$
35 Phosphohypotaurocyamine	$H_2O_3PNH-\overset{\overset{\displaystyle NH}{\|\|}}{C}-NH-(CH_2)_2-SO_2H$
36 Phosphoopheline	$H_2O_3PNH-\overset{\overset{\displaystyle NH}{\|\|}}{C}-NH-(CH_2)_2-O\overset{\overset{\displaystyle O}{\|\|}}{\underset{\underset{\displaystyle OH}{\|}}{P}}-OMe$
37 Phospholombricine	$H_2O_3PNH-\overset{\overset{\displaystyle NH}{\|\|}}{C}-NH-(CH_2)_2-O\overset{\overset{\displaystyle O}{\|\|}}{\underset{\underset{\displaystyle OH}{\|}}{P}}-OCH_2-\underset{\underset{\displaystyle COOH}{\|}}{CH}-NH_2$
38 Phosphothalassemine	$H_2O_3PNH-\overset{\overset{\displaystyle NH}{\|\|}}{C}-NH-(CH_2)_2-O\overset{\overset{\displaystyle O}{\|\|}}{\underset{\underset{\displaystyle OH}{\|}}{P}}-OCH_2-\underset{\underset{\displaystyle COOH}{\|}}{CH}-NMe_2$

during such activity. Under these circumstances, the phosphagen system offers several advantages over a hypothetical alternative system characterized by a high ATP concentration (Watts, 1977): (1) avoidance of the phosphate loss; (2) low concentration of counterions; (3) lower production of ADP, a regulator of various enzymatic reactions; and (4) no acidification by hydrolysis of ATP.

The tissue phosphagen concentration reflects the role of these compounds. Phosphagen concentration is considerable in muscles, especially those which must provide intense work. For instance, this concentration is 33.36 μmoles/g wet weight in the muscle of *Homarus vulgaris* and 52.16

μmoles/g wet weight in the adductor muscle of the scallop *Pecten maximus* (Beis and Newsholme, 1975). In both cases, rapid contractions of these muscles result in the swimming movements of the escape reaction. This short but intense effort requires considerable energy; hence, the phosphagen concentration is quite high. The swimming movements of fish are much less intense, and, correspondingly, the observed concentration in *Scylliorhinus canicula* muscle is only 8.8 μmoles/g wet weight (Beis and Newsholme, 1975). The phosphagen concentration in other tissues has been less well studied, but in general it is much lower than that of muscle, approximately ten times lower in ovocytes and three or four times lower in spermatozoa (Thoai and Robin, 1969).

The distribution of phosphagens among marine taxonomic groups displays some interesting features. Phosphocreatine (**32**) is the phosphagen of vertebrates, while phosphoarginine (**31**) seems to be typical of invertebrates. However, in the latter group both are sometimes present as shown in Table 2. The phosphagen pattern of marine worms is quite interesting, since in addition to phosphoarginine and phosphocreatine, six novel phosphagens have been found, five of which are found only in marine worms. Different species of the same genus may even contain different phosphagens. For example, in the genus *Ophelia* (segmented worms),

TABLE 2

Phosphagens and Biochemical Evolution

Animal origin	Phosphagens[a]							
	31	**32**	**33**	**34**	**35**	**36**	**37**	**38**
Bacteria	+							
Protozoa	+							
Porifera	+	+						
Coelenterata	+							
Nemathelminthes	+							
Nemertinea	+							
Platyhelminthes	+							
Annelida, Echiura, and Sipunculida	+	+	+	+	+	+	+	+
Arthropoda	+	Dead-end mutations						
Mollusca	+							
Echinodermata	+	+						
Protochordata	+	+						
Vertebrata		+						
	Normal evolutionary pathway							

[a] This table does not take into account phosphagens of gametes. Adapted from "Phosphagens and molecular evolution in worms" (Robin, 1974) with the permission of Y. Robin and Elsevier.

phosphoopheline (**36**) is found in *O. neglecta,* phospholombricine (**37**) in *O. bicornis,* and phosphotaurocyamine (**34**) in *O. radiata* (Thoai *et al.,* 1963b).

The distribution of phosphagens in tissues is not always straightforward. In marine invertebrates, different phosphagens are frequently found in different tissues of the same animal. For instance, in *O. neglecta,* phosphoopheline (**36**) is present in muscles, phospholombricine (**37**) in ovocytes, and phosphocreatine (**32**) in spermatozoa (Thoai and Robin, 1969). A similar pattern is observed for *Thalassema neptuni,* but in this case thalassemine takes the place of opheline (Thoai *et al.,* 1972). In addition, different phosphagens are sometimes found even within the same tissue. However, this observation has been made only in some marine invertebrates, specifically, sea urchins and sea worms. In a detailed study concerning phosphagen composition in polychaetes and sipunculids collected in California and in Brittany, Robin (1964a,c) showed that a great number of the animals contained two and even three phosphagens.

Attempts have been made to study the muscular evolution and phylogeny of the animal kingdom using phosphagens as markers; however, the results obtained for the past few decades indicate that the picture is actually a great deal more complex than had been thought. Nevertheless, some general relationships between phylogeny and phosphagen composition may be demonstrated as shown in Table 2. The most primitive invertebrates possess only phosphoarginine; two more highly evolved phyla Echinodermata and Protochordata, contain both phosphoarginine and phosphocreatine, whereas vertebrates possess only the latter. However, there exist some unexplained important exceptions: the primitive phylum Porifera contains both, and the phyla Annelida, Sipunculida, and Echiura display an outstanding diversity of phosphagen composition. In a more recent attempt at chemotaxonomy, phosphagen phosphotransferases have been used as markers. In that case, an obvious homology exists between different phosphagen kinases, which reflects evolutionary relationships much more closely (Robin, 1974 and cited references).

III. ACARNIDINES AND POLYANDROCARPIDINES

As Faulkner (1977) has pointed out, "often many marine organisms appear to contain mainly fatty acid derived lipids." Such compounds are often discarded without appropriate study; however, recent work has shown that some fatty acid derivatives display interesting structural fea-

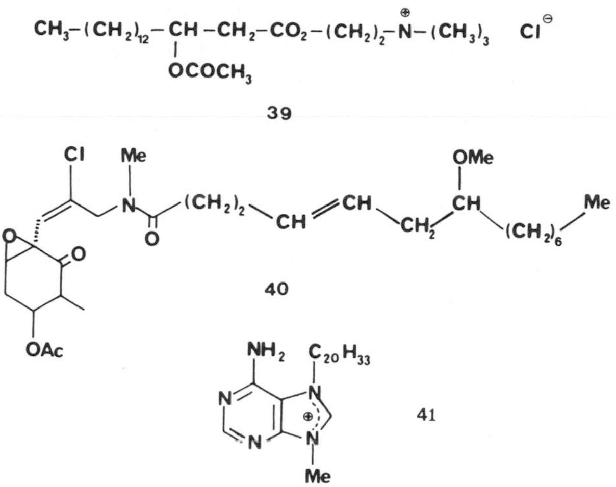

$$CH_3-(CH_2)_{12}-CH-CH_2-CO_2-(CH_2)_2-\overset{\oplus}{N}-(CH_3)_3 \quad Cl^{\ominus}$$
$$| \\ OCOCH_3$$

39

40

41

Fig. 2. Examples of marine nitrogenous long-chain derivatives.

tures. This is especially true with respect to nitrogenous derivatives of fatty acids. Such compounds have been found in various phyla, notably pahutoxin (39) from fish (Boylan and Scheuer, 1967), stylocheilamide (40) from Mollusca (Rose, 1975; Rose et al., 1978), and agelasine (41) from Porifera (Cullen and Delvin, 1975). The formulas are shown in Fig. 2.

Two new series of compounds found by Rinehart and co-workers may be placed in this class: acarnidines (42a,b,c) and polyandrocarpidines (43a,b), both containing a guanidino group. The $C_{12:0}$, $C_{12:1}$, and $C_{14:3}$ acarnidines (42a,b,c) were isolated from an extract of the sponge *Acarnus erithacus* (DeLaubenfels) collected in the Gulf of California (Carter and Rinehart, 1978). The chemical structures were established by spectral analysis of 42a,b,c and of the 4,6-dimethyl pyrimidine derivatives (44a,b,c, 45a,b) obtained by treatment of 42a,b,c and their hydrogenated derivatives with 2,4-pentanedione. Synthesis of 45b, following the scheme shown in Fig. 3, confirmed the backbone structure of 42c. Double-bond positions in the fatty acid residues were located by ozonolysis of the acarnidine mixture. Recently, the structure of $C_{14:3}$ acarnidine (42c) was slightly revised from the previously proposed formula to that shown in Fig. 3 (Rinehart, personal communication). Acarnidines show significant inhibitory activity against *Herpes simplex* virus and against various microorganisms, such as *Bacillus subtilis* and *Penicillium atrovenetum*.

The structurally related compounds polyandrocarpidines I and II (43a,b) (Fig. 5) are produced by a very different organism, *Polyandrocarpa sp.*, a colonial tunicate (Cheng and Rinehart, 1978). A remarkable feature of these compounds is the presence of a cyclopropene ring.

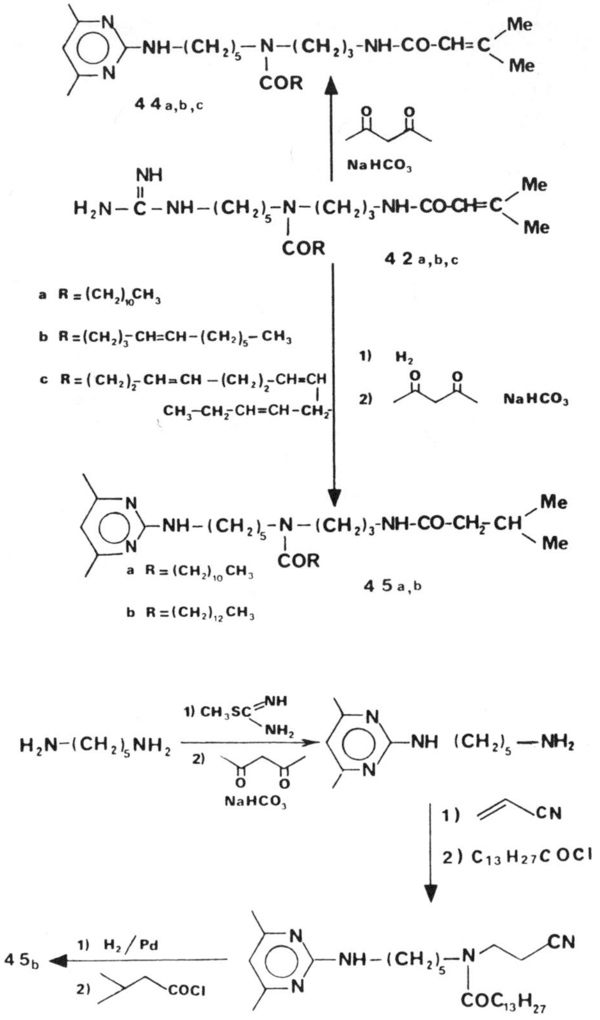

Fig. 3. Derivation and synthesis of acarnidines.

This functional group has previously been encountered only twice in the field of natural products. Calysterol (**46**), a sterol isolated from a sponge, possesses a cyclopropene ring in the side chain (Fattorusso *et al.*, 1975) while malvalic (**47**), sterculic (**48**), and related acids are fatty acids which also contain this strained ring (Devon and Scott, 1975) (see Fig. 4). The structures of the polyandrocarpidines were deduced mainly by extensive studies of mass spectroscopic data of the pyrimidine derivatives **49a,b** and **50a,b** (see Fig. 5). Synthesis of **49a** has confirmed the formula proposed for **43a.** Mixtures of the polyandrocarpidines display some interesting

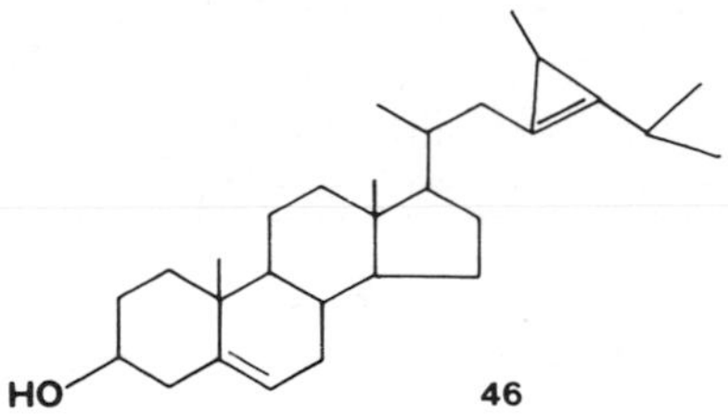

46

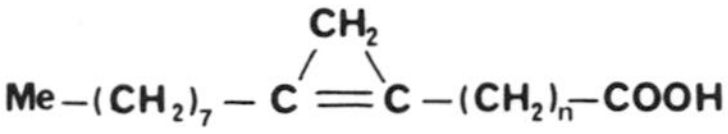

$$Me-(CH_2)_7 - C \overset{\displaystyle CH_2}{=} C -(CH_2)_n-COOH$$

47 $n = 6$

48 $n = 7$

Fig. 4. Natural cyclopropene-containing products.

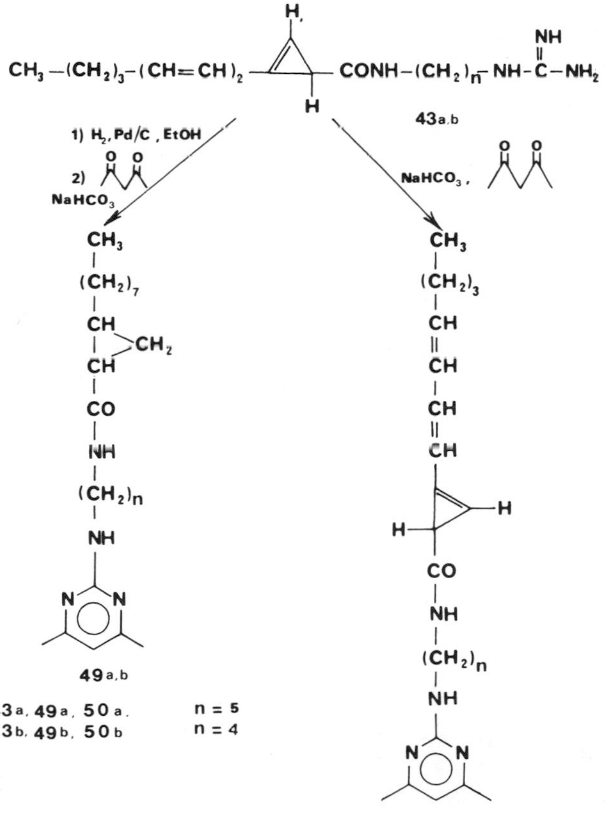

$$CH_3-(CH_2)_3-(CH=CH)_2-\text{[cyclopropene]}-CONH-(CH_2)_n-NH-\overset{NH}{\underset{}{C}}-NH_2$$

43a,b

1) H_2, Pd/C , EtOH

2) NaHCO$_3$

NaHCO$_3$,

CH$_3$
|
(CH$_2$)$_7$
|
CH
| >CH$_2$
CH
|
CO
|
NH
|
(CH$_2$)$_n$
|
NH
|
[pyrimidine]

49a,b

43a, 49a, 50a.   n = 5
43b, 49b, 50b   n = 4

CH$_3$
|
(CH$_2$)$_3$
|
CH
||
CH
|
CH
||
CH
|
[cyclopropene]
|
CO
|
NH
|
(CH$_2$)$_n$
|
NH
|
[pyrimidine]

50a,b

Fig. 5. Polyandrocarpidines and derivatives.

pharmacological properties: bactericidal or bacteriostatic activity against *Bacillus subtilis, Staphylococcus aureus,* and *Mycobacterium avium* and antiviral activity against *Herpes* virus. The mixtures are also cytotoxic.

IV. OROIDIN AND RELATED COMPOUNDS

A series of intriguing compounds which contain a bromopyrrole moiety with a guanidine residue linked in different ways has been isolated from various sponges. The most representative product in this series is probably oroidin (**51**) which was isolated from a butanol extract of *Agelas oroides* collected in the bay of Naples (Forenza *et al.*, 1971). The Italian researchers proposed structure **52** for oroidin, since after basic hydrolysis they obtained amide **53** as a by-product, in addition to acid **54** (see Fig. 6). They therefore concluded that the nitrogen atom of the carboxamide unit is linked to a hydrolyzable group such as guanidine. In fact, Garcia *et al.* (1973) corrected the structure of **51** following synthesis. In addition, *Agelas oroides* elaborates the bromopyrroles **53, 54,** and **55**. Oroidin has also been found in two sponges of the genus *Axinella, A. damicornis* and *A. verrucosa* (Cimino *et al.*, 1975).

Some months prior to the Italian work, Sharma and Burkholder (1971) had obtained 4-bromophakellin (**56a**) and dibromophakellin (**56b**) from *Phakellia flabellata,* a sponge collected on the Great Barrier Reef. These compounds show some antibiotic activity (Sharma *et al.,* 1970). The structure of **56b** was established by interpretation of ^{1}H nmr spectra of the phakellins and their derivatives. It has also been confirmed by single crystall X-ray diffraction analysis of the acetyl derivative (**56c**). Recently

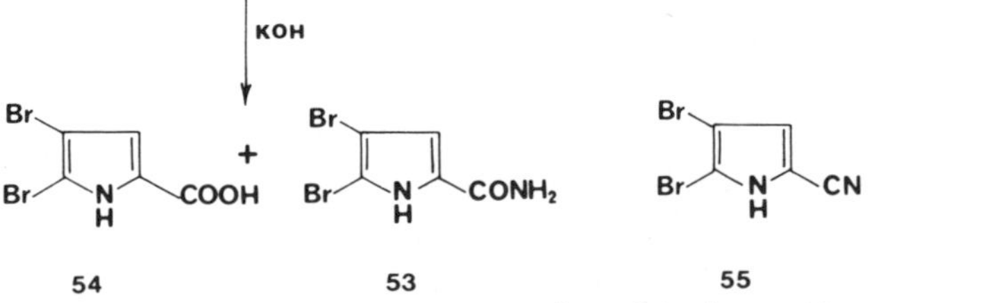

Fig. 6.   Nitrogenous metabolites of *Agelas oroides.*

Baker (1976) reported that an alternate tricyclic structure (**57**) could be consistent with 220-MHz ^{1}H nmr spectroscopic data, and that cyclization might occur during acetylation. Sharma and Burkholder (1971) also pointed out the unusually low basicity ($pK_a = 7.7$) of the guanidine residue of this compound. Shimizu (1978), drew attention to the strong structural similarity between the phakellins (**56**) and saxitoxin (**58**) in which the second pK_a value (8.24) is also low.

Sharma and Maydoff-Fairchild (1977) proposed an explanation for the low basicity of phakellins. The strong basicity of guanidines is due largely to the gain of resonance energy accompanying the formation of the guanidinium ion; therefore, to obtain this stabilization, all three nitrogen atoms must adopt a planar arrangement, in such a way that all three nitrogen electronic doublets are parallel. In the case of the phakellins (**56**), however, this may only be achieved if the imidazoline ring becomes planar. In fact, X-ray diffraction studies show that the imidazoline ring of (**56c**) is twisted in the free base form. In addition, an inspection of molecular models proves that such a requirement is difficult to satisfy because it produces strong conformational strains in the central six-membered ring. For this reason, the phakellins do not display the usual strong basicity of guanidines. This hypothesis is satisfying as such, al-

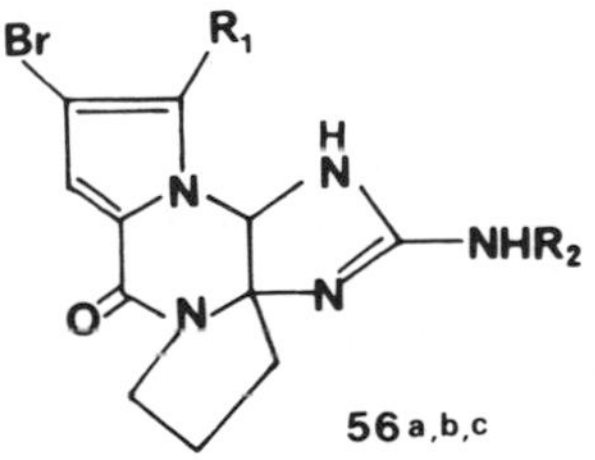

56 a,b,c

a $R_1 = H$ $R_2 = H$
b $R_1 = Br$ $R_2 = H$
c $R_1 = Br$ $R_2 = COCH_3$

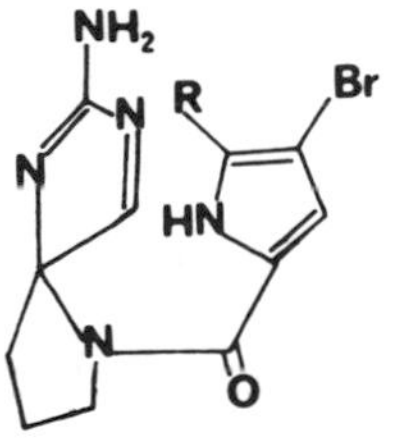

57

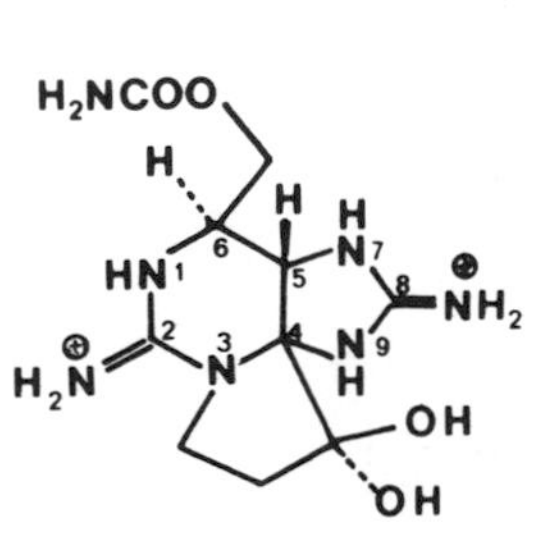

58

though only indirect evidence has been given. Obviously, a similar explanation may be advanced to explain the low basicity of the second guanidine function of saxitoxin (58), since in the p-bromobenzene-sulfonate saxitoxin, N—1, C—2, N—3, C—4, and C—6 are not coplanar as shown by X-ray diffraction studies (Schantz et al., 1975; vide infra). It should be noted in passing that the strong seasonally dependent antibiotic activity of the methanol extract of Phakellia ventilabrum is not due to the weak antibacterial power of phakellins, but to some other product(s).

Some years later, Scheuer's group isolated a related compound, midpacamide (59) from a sponge, Agelas cf mauritiana, collected on Enewetak atoll (Chevolot et al., 1977). The structural similarity with oroidin is strong, with the exception that a methylhydantoin residue replaces the aminoimidazole ring. Structural determination was effected by spectral analysis and by study of the hydrolysis products. The major secondary metabolite of this sponge is the 4,5-dibromo-1-methyl-2-pyrrolecarboxylic acid (60).

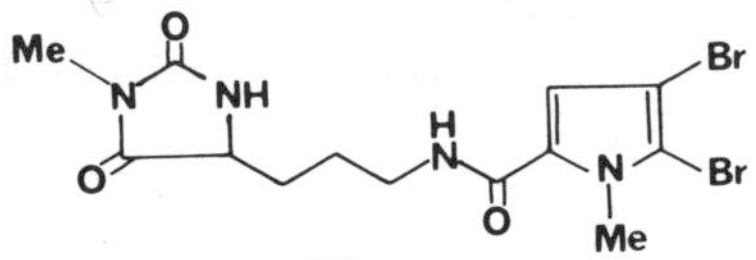

59

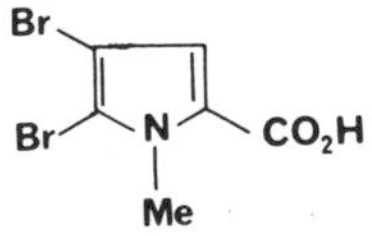

60

Obviously, a strong biogenetic similarity exists among oroidin, phakellin, and midcapamide. The pyrrole unit could be derived from proline and the other moiety from ornithine, which could be amidinated on the α-NH$_2$. N-Amidino-4-bromopyrrole-2-carboxamide (61) was isolated from a sponge of the genus Agelas (Stempien et al., 1972); this compound displays bactericidal activity.

This group of compounds as well as others (such as sterols) aid in the classification of Desmospongiae. Cimino et al. (1975) used such compounds for this purpose. They concluded that Agelas (order

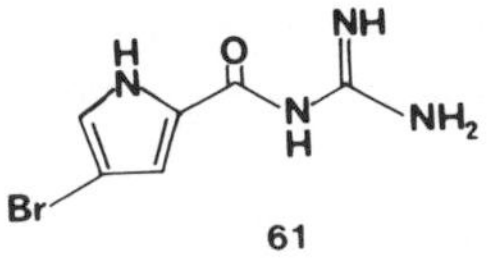

61

Paecilosclerida) and at least some *Phakellia* and *Axinella* spp. (family Axinellidae, order Axinellida) are biochemically related because they contain oroidin-like substances and/or bromopyrroles. Bergquist and Hartmann (1969) found that sponges of the genus *Agelas* display a free amino acid pattern quite atypical of the order Paecilosclerida. On the other hand, based on the same criterion, some Axinellidae appear to be related to the order Paecilosclerida. In fact, the chemical composition of the Axinellidae seems heterogeneous both with respect to amino acids or secondary metabolites (unusual sterols, isonitrile terpenes, bromopyrroles, etc.). Up to the present time, however, the number of samples studied is insufficient to permit definite conclusions to be drawn.

V. APLYSINOPSINE

Aplysinopsine (**62b**) was discovered almost simultaneously by several Australian researchers in some Great Barrier Reef sponges of the genus *Thorecta* (Kazlauskas *et al.*, 1977) and also by Hollenbeak and Schmitz (1977) in *Verongia spengelii*, a sponge collected at Sheriff Key, Florida. In addition, Kazlauskas and co-workers also detected small quantities (5%) of an aplysinopsine bromo analog in the mass spectrum. Derivative **62a** was also obtained as a major product, but this was probably an artifact. Some structural elements have been identified by spectroscopic methods. Acetylation of **62a** followed by reduction with Zn/AcOH gave three compounds, of which the structure of product **63** was deduced by single crystal X-ray diffraction. The structures of aplysinopsine (**62b**) and the derivative **62a** were thus established. Synthesis of aplysinopsine by condensation of **64** and **65** confirmed this result (see Fig. 7). The Australian workers did not specify the stereochemistry of the double bond, but they stated that natural and synthetic aplysinopsines are mixtures (10:1) of two stereoisomers. The American group (Hollenbeak and Schmitz, 1977) determined the configuration of aplysinopsine diacetate by measuring significant nuclear Overhauser effects between the N–Me and the olefinic proton on one hand, and between the indole 2-hydrogen atom and the olefinic proton on the other hand. Hence, at least for this derivative, the major stereoisomer has the configuration shown in Fig. 7. The stereochemistry of the companion molecule (12%) has not been determined.

American researchers showed that aplysinopsine is responsible for the antitumor activity of the crude extract of *V. spengelii*. The aplysinopsine ED_{50} (effective dose which inhibits 50% of cell growth) are 0.87 μg/ml, 3.8 μg/ml, and 3.7 μg/ml against KB, P-388, and L-1210 cells, respectively. In

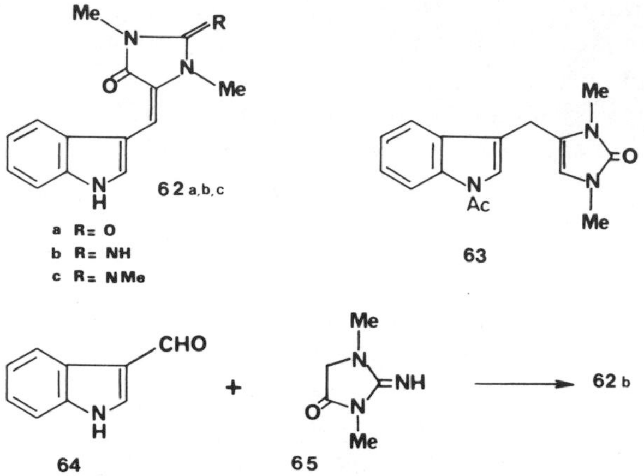

Fig. 7. Outline of synthesis of aplysinopsine and derivatives.

addition, methylaplysinopsine (62c) has also been isolated from a marine sponge. This is probably the first substance with commercial potential, according to the RRIMP* group. This compound is an interesting antidepressant, acting as a competitive inhibitor of monoamine oxidase (Taylor *et al.*, 1978).

These findings are of interest from the chemotaxonomic point of view, since *Thorecta* and *Verongia* are placed in two different orders, Dictyoceratida and Verongida, respectively, in modern classification (Bergquist, 1978), whereas, according to Hollenbeak and Schmitz (1977), De Laubenfels (1948) placed these sponges in the same subfamily. Furthermore, the Australian workers (Kazlauskas *et al.*, 1977) reported that Von Lendenfels (1889) distinguished two genera (*Thorecta* and *Aplysinopsis*) in place of a single genus *Thorecta*. These authors added that sponges resembling *Aplysinopsis* contain aplysinopsine, while sponges resembling *Thorecta* produce sesterterpenes.

VI. ZOANTHOXANTHINS

Recently, an Italian group (Prota and co-workers) has isolated a series of novel yellow pigments which are highly fluorescent under ordinary light. Moreover, Prota (1979) has reviewed this kind of new nitrogen-containing tricyclic metabolites which possess either the 1,3,5,7-tetrazaclopent(*f*)azulene skeleton (66) or the 1,3,7,9-tetrazacyclopent(*e*)azulene

*Roche Research Institute of Marine Pharmacology. Dee Why N.S.W. (Australia).

66   67

structure (67). Until now, this type of compound has been isolated only from colonial anthozoans of the order Zoanthidea (subclass Zoantharia).

The first of these compounds to be isolated containing the 66 skeleton was zoanthoxanthin (68) (Cariello *et al.*, 1973, 1974a), from a subspecies of *Parazoanthus axinellae,* tentatively identified as *P. a. adriaticus.* Purification of zoanthoxanthin was accomplished by solvent partitioning and passage through a cation-exchange resin, followed by recrystallization from methanol. Chemical studies and spectroscopic data provided some information about the molecule; but since zoanthoxanthin is stable toward the usual degradative reagents, a more complete study of the molecule by such methods (i.e., an analysis of degradation products) was impracticable. The zoanthoxanthin structure was fully determined by X-ray diffraction analysis of the chloro- analog (69) obtained by diazotation in hydrochloric acid medium. The authors checked the integrity of the initial structure during diazotation by regenerating zoanthoxanthin with ammonia from the chloro derivative (69). In addition, X-ray analysis showed that the molecule is not strictly planar. The central ring adopts a shallow boat conformation, as in other seven-membered ring systems. Seven additional related pigments, parazoanthoxanthins A, B, C, D, E, F, and G (70–76) have been obtained from the same species (Cariello *et al.,* 1974b,c, 1979). Their structures were determined spectroscopically and by correlation with synthetic samples. The Italian workers obtained two additional zoanthoxanthins with the same skeleton from the related species *Parazoanthus axinellae* (originally considered *Epizoanthus arenaceus*). The structures of these zoanthoxanthins epizoanthoxanthin A (77) (Cariello *et al.,* 1974c,d) and epizoanthoxanthin B (78) (Cariello *et al.,* 1974d) were determined using the same methods. In a subsequent study, palyzoanthoxanthins A, B, and C (79–81) were obtained from *Palythoa mammilosa* or *P. tuberculosa* in minute quantities (Cariello *et al.,* 1979). The structures are shown in Table 3.

In the second structural group the first compounds isolated were pseudozoanthoxanthin (82) and 3-norpseudozoanthoxanthin (83) from *Parazoanthus axinellae* (Cariello *et al.,* 1974d). After analysis of the spectroscopic and chemical data, three possible structures were postulated for pseudozoanthoxanthin: 82, 84, and 85 (formulas shown in Tables

TABLE 3

1,3,5,7-Tetrazacyclopent (f)azulene Pigments from Zoanthids[a]

Pigment	R_1	R_2	R_3	R_6
68 Zoanthoxanthin	—	NH_2	Me	NMe_2
69 Chloro derivative	—	Cl	Me	NMe_2
70 Parazoanthoxanthin A	H	NH_2	—	NH_2
71 Parazoanthoxanthin B	—	NH_2	Me	NH_2
73 Parazoanthoxanthin D	H	NH_2	—	NMe_2
74 Parazoanthoxanthin E	—	NHMe	Me	NMe_2
75 Parazoanthoxanthin F	Me	=NH	Me	NMe_2
76 Parazoanthoxanthin G	—	NH_2	Me	NHMe
77 Epizoanthoxanthin A	H	NHMe	—	NMe_2
78 Epizoanthoxanthin B	Me	NMe_2	—	NHMe
79 Palyzoanthoxanthin A	Me	NH_2	—	NMe_2
80 Palyzoanthoxanthin B	H	NMe_2	—	NMe_2
81 Palyzoanthoxanthin C	—	NMe_2	Me	NMe_2
84 Synthetic product	Me	=NH	Me	NH_2

[a] Radical numbering is consistent with the cyclic numbering of Cariello *et al.* (1974a). When R_1 or R_3 = H, the H is arbitrarily placed at N-1.

3 and 4). Structure **84** was ruled out by direct comparison of the natural product with the semisynthetic compound possessing structure **84**. Lack of an NOE between C–Me and N–7-Me eliminated **85**. The structure **82** is therefore assigned to pseudozoanthoxanthin. Demethylation of **82** with boiling concentrated hydrobromic acid gave a compound identical with the natural 3-norpseudozoanthoxanthin (**83**). In fact, there are two types of compounds in the pseudozoanthoxanthin series. Compounds **82** and **83** belong to Type 1 (see Table 4). Scheuer's group isolated the first two compounds (**86** and **87**) of Type 2 from an Hawaiian *Gerardia* sp. known as gold coral, prized for the manufacture of jewelry. The structure of **86** was found to be planar and was determined by single crystal X-ray techniques (Schwartz *et al.*, 1978, 1979), while that of **87** was determined by chemical correlation (Schwartz *et al.*, 1979).

A group of Japanese workers also obtained a pseudozoanthoxanthin product from the anthozoan *Parazoanthus gracilis*. They named it paragracine (**88**) and determined its structure by X-ray crystallographic analysis of the dihydrobromide salt (Komoda *et al.*, 1975). Six additional compounds belonging to Type 1 (**82**), Type 2 (**86, 91, 92**), and both **89** and **90**

TABLE 4

1,3,7,9-Tetracyclopent(e)azulene Pigments from Zoanthids[a]

Type 1   or   Type 2

Pigment		Type	R_1	R_2	R_3	R_7	R_8	R_9
82	Pseudozoanthoxanthin	1	Me	$=NH$	Me	—	NH_2	—
83	3-Norpseudozoanthoxanthin	1	Me	NH_2	—	—	NH_2	—
85		2	—	NH_2	—	Me	$=NH$	Me
86		2	—	NMe_2	—	—	NH_2	Me
87		2	—	$NHMe$	—	—	NH_2	Me
88	Paragracine		H	NMe_2	—	—	$NHMe$	—
89	Norparagracine		H	$NHMe$	—	—	$NHMe$	—
90			H	NMe_2	—	—	NH_2	—
91		2	—	NMe_2	—	—	$NHMe$	Me
92		2	—	NH_2	—	—	NH_2	Me
95			H	NH_2	—	—	NH_2	—

[a] Radical numbering is consistent with the cyclic numbering of Cariello et al. (1974d). The methylation patterns at nuclear nitrogen distinguish between Type 1 and Type 2; when there is only H as a substituent, both types are identical and the H is placed arbitrarily at N-1.

were subsequently isolated from the same source (Komoda, personal communication).

Cariello et al. (1979) searched for zoanthoxanthins in a large number of marine invertebrates. They detected such compounds only in zoanthids, and only found relatively large quantities in the genus Parazoanthus. It must be pointed out that the presence of large quantities of 86 and 87 in Hawaiian Gerardia is inconsistent with this result (vide supra). These compounds possess interesting biological activities. For example, paragracine displays papaverine-like pharmacological properties (Komoda et al., 1975). In addition, zoanthoxanthin (68) and 3-norzoanthoxanthin (73) are DNA intercalating agents. Both compounds selectively inhibit DNA synthesis, but zoanthoxanthin (68) is more effective than is 73 (Quadrifoglio et al., 1975). From a biogenetic point of view, Cariello et al. (1974b) proposed that these compounds result from the duplication of two C_5N_3 units derived from arginine. There is no direct evidence to support this hypothesis, but Braun and Büchi (1976) synthesized both skeletons by dimerization of 93 in concentrated sulfuric acid following a biomimetic pathway. In fact, products 70 and 95 resulted from the (6 + 4)-

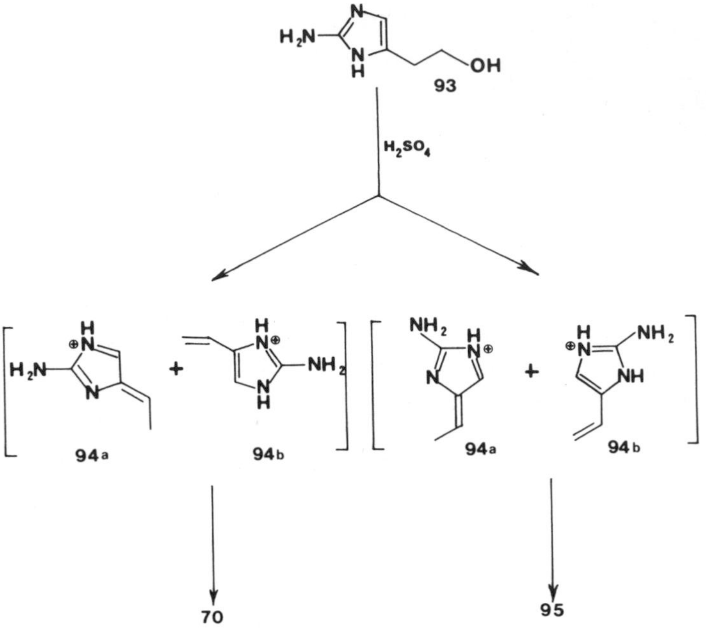

L. Chevolot

Fig. 8. Synthesis of zoanthoxanthins.

cycloaddition of **94a** and **94b** in the two ways indicated in Fig. 8. However, formation of **70** by (6 + 4)-cycloaddition appeared doubtful according to Braun *et al.* (1978). In fact, the ratio of **70** to **95** depends on starting materials and reaction conditions. From these two products, most zoanthoxanthins may be synthesized by specific methylation processes described by Cariello *et al.* (1974c). In neutral or acidic medium, methylation occurs first at N-1 and subsequently at N-3. In basic medium, only the primary amino groups are methylated.

The nomenclature in this field obviously lacks homogeneity. It is preferable to follow the suggestion of Schwartz *et al.* (1979): Compounds possessing skeleton **66** or **67** are called, respectively, zoanthoxanthin and pseudozoanthoxanthin with a consistent numbering. The position number of saturated nuclear nitrogen followed with H distinguishes between the various groups of zoanthoxanthins.

VII. TETRODOTOXIN AND SAXITOXINS

Tetrodotoxin (**96**) and saxitoxin (**58**) are the two most important marine guanidino compounds, both with respect to their outstanding structural features and to public health problems. Tetrodotoxin (96) is found frequently in the viscera, skin, and ovaries of some species of the Tetradontidae. This type of fish is highly prized by Japanese gourmets, and its consumption has posed a difficult problem to Japanese authorities. The biological origin of this compound remains unknown. Some evidence, such as maximum toxicity of the fish during spawning months, points to an endogenous origin; however, even during that period no more than 50% of the fish are toxic. This toxin is also present in the unrelated fish *Gobius criniger* (Gobiidae) (Noguchi and Hashimoto, 1973), in the Californian salamander *Taricha torosa* (Mosher *et al.*, 1964), in the octopus *Hapalochlaena maculosa* (Scheumack *et al.*, 1978), and in the skin and eggs of the Costa Rican frog *Atelopus chiriquiensis* (Pavelka *et al.*, 1977). In this last organism, at least one other toxin, chiriquitoxin, is also present. Chiriquitoxin appears to be similar to tetrodotoxin. In *A. zeteki*, some guanidino group-containing toxins are present, but tetrodotoxin is absent (Brown *et al.*, 1977).

96

The molecular structure of tetrodotoxin (96) is characterized by a guanidino group and a unique hemilactal function. As with many marine toxins, the ratio of heteroatoms to carbon is very high: $C_{11}H_{17}N_3O_8$ is the gross formula of tetrodotoxin. Its structure was elucidated in 1964 almost simultaneously in two laboratories in Japan and two in the United States (Scheuer, 1973). Total synthesis was achieved in a remarkable manner by Kishi and co-workers (Kishi *et al.*, 1972; Tanino *et al.*, 1974). The reader is referred to Scheuer (1973) for previous synthesic work.

Saxitoxin (58) was first isolated from the giant clam *Saxidomus giganteus,* but in fact it is synthesized by the dinoflagellate *Gonyaulax catenella,* the etiologic agent of this toxin in the Pacific ocean. During blooms of this species, which are known as red tide, the toxin is concentrated in shellfish and causes the phenomenon known as paralytic shellfish poisoning (PSP). In the North Atlantic ocean, *G. tamarensis* is the organism responsible for PSP; however, this dinoflagellate produces little saxitoxin, but chiefly neosaxitoxin and gonyautoxins II and III (GTX II

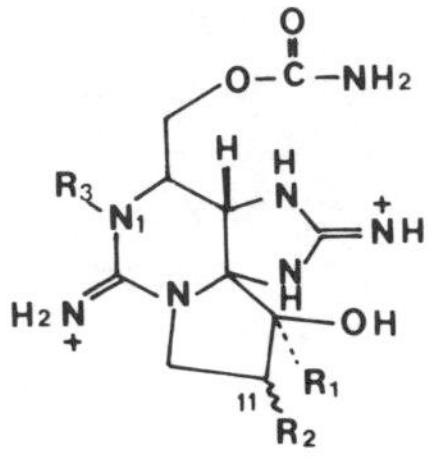

58	R1 = OH	R2 = H	R3 = H
97	R1 = OH	R2 = H	R3 = OH
98	R1 = H	R2 = H	R3 = H
100	R1 = OH	R2 = αOH	R3 = H
101	R1 = OH	R2 = βOH	R3 = H
102	R1 = OH	R2 = OSO3H	R3 = H
103	R1 = OH	R2 = αOSO3H	R3 = H
104	R1 = OH	R2 = βOSO3H3	R3 = H

Fig. 9. Saxitoxin and related compounds.

and III). Saxitoxin is also synthesized by a freshwater blue-green alga, *Aphanizomenon flos-aquae,* together with three other related toxins (Moore, 1977). Interestingly, saxitoxin is found in the crab *Zosimus aeneus,* mainly in the exoskeleton (Noguchi *et al.*, 1969; Scheuer, 1977).

The structural elucidation of saxitoxin (**58**) posed a very difficult problem for chemists (Shimizu, 1978). It was finally solved by X-ray diffraction techniques on the crystalline salts of saxitoxin (Schantz *et al.*, 1975; Bordner *et al.*, 1975) (see Fig. 9).

Shimizu *et al.* (1978b) have recently established the structure of neosaxitoxin as *N*-1-hydroxysaxitoxin (**97**). This structure has been assigned following careful examination of ^{13}C nmr spectra and by chemical transformation of neosaxitoxin (**97**) to saxitoxin (**58**) and dihydrosaxitoxin (**98**) by zinc–acetic acid or sodium borohydride reduction. Treatment of **97** with pyridine/acetic anhydride gives a uv-absorbing (295 nm) product with the tentative structure **99**. This reaction is similar to the Polonovski reaction, well known in the field of alkaloid chemistry (Potier, 1978).

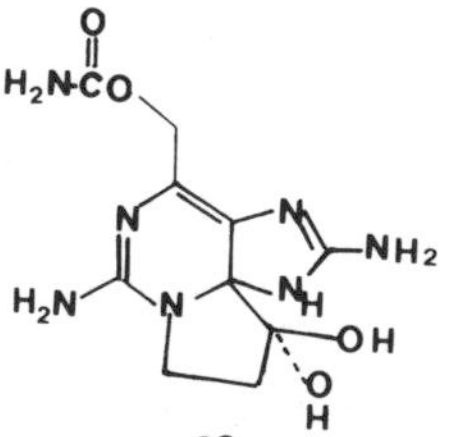

99

Shimizu *et al.* (1976) deduced the structures of GTX II and III as 11α-hydroxysaxitoxin (**100**) and 11β-hydroxysaxitoxin (**101**), respectively, on the basis of the 44 ppm downfield shift of C-11 in the ^{13}C nmr spectra. Chemical techniques gave further support for this structure, but no molecular-weight determination was performed. More recently, a sulfate ester of 11-hydroxysaxitoxin (**102**) was isolated from scallops infested by *G. tamarensis* (Boyer *et al.*, 1978). The presence of a sulfate residue was suggested by the electrophoretic mobility of **102** and confirmed by chemical analysis. Since **102** and GTX II and III all display similar electrophoretic behavior, it was suggested that the structures of GTX II and III are in fact **103** and **104**, rather than **100** and **101**. The first total synthesis of saxitoxin was reported as recently as 1977 (Tanino *et al.*, 1977).

Tetrodotoxin, saxitoxin, and other similar compounds are powerful neurotoxins which inhibit sodium passage through axonal membranes. The presence of a positive charge on the guanidinium ion is probably an important feature in blocking the sodium channels. Increased knowledge of these compounds is thus important to the understanding of sodium transport mechanisms. In this respect, it should be noted that coldblooded organisms are much less sensitive toward these toxins than warmblooded. Some complete and recent reviews have been published concerning marine toxins including tetrodotoxin and saxitoxin (Baslow, 1971; Scheuer, 1975, 1977). Shimzu (1978) has completely reviewed saxitoxin and other dinoflagellate toxins. For details concerning the physiological properties of saxitoxin (**58**) and tetrodotoxin (**96**), the reader is referred to Kao (1972), Narahashi (1972), Hille (1975), Blankenship (1976), Ulbricht (1977), and Ritchie (1978).

Little is known about the biogenesis of these toxins; however, both compounds, tetrodotoxin and saxitoxin, are probably derived from arginine. To form tetrodotoxin (**96**), an additional C_5 unit is required, which could be derived from valine, leucine, or mevalonic acid. It is well known in the field of indole alkaloid chemistry that the C_9-C_{10} unit is derived from terpenes. Saxitoxin (**58**), on the other hand, could be built from a guanidinoarginine and a C_2 unit (see Fig. 10), but it should be also derived from the purine secondary metabolism.

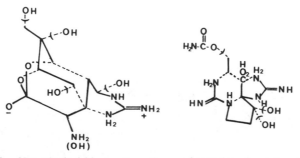

Fig. 10.   Hypothetical biogenetic pathway of tetrodotoxin and saxitoxin.

VIII. OTHER CYCLIC COMPOUNDS

A. Guanine Derivatives

Guanine and its common derivatives, as well as other purine bases, are widely distributed among marine organisms. However, some nonclassical compounds have also been isolated from various marine organisms. Ackermann and List (1957, 1960) isolated a new guanidinium betaine, herbipoline (**105**), from the sponge *Geodia gigas*. From another sample of the same sponge collected in Greece, they found spongopurine (**106**) although they were unable to reisolate herbipoline (Ackermann and List, 1961).

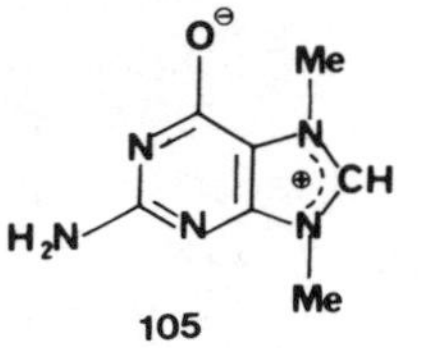

105 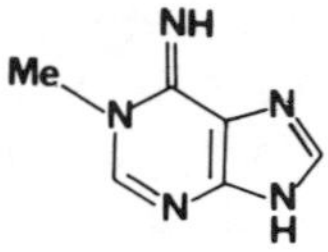   106

From an ascidian, *Microcosmus polymorphus,* 2-amino-6,8-dioxypurine (**107**) has been isolated by hot-water extraction (Karrer *et al.*, 1948; Ciereszko, 1970).

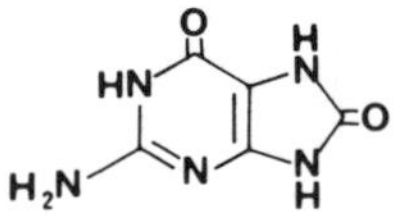

107

B. Pterins

Pteridines are derivatives of tetrazanaphthalene (**108**) and constitute a class of natural compounds. The English biochemist Hopkins isolated the first compounds in this series from butterfly wings.

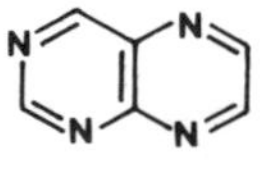

108

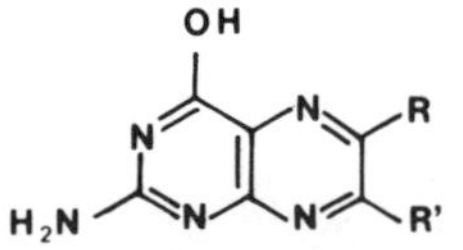

109 a	R = H	R' = H
109 b	R = OH	R' = H
109 c	R = CHOH-CHOH-CH₂OH	R' = H
109 d	R = H	R' = OH
109 e	R = CHOH-CHOH-CH₃	R' = OH
109 f	R = COOH	R' = OH

These types of compounds have been recently reviewed (Ziegler and Harmsen, 1969; Pfleiderer, 1975; Kisliuk and Brown, 1979). Pterins are usually defined as 2-amino-4-hydroxypteridines (109a). These compounds play various important metabolic roles: hydroxylation, electron transport, C_1 unit transfer. (It may be recalled that folic acid contains a such unit.) In addition, pterins are also pigments frequently found in the epidermis and eyes of insects (Corrigan, 1970) and in the skin of amphibians (Balinski, 1970).

In the marine environment, xanthopterin (109b) and other pteridine pigments are frequently present in crustaceans (Goodwin, 1971) and more seldom in fishes (Lee *et al.*, 1969; Liaci, 1970). However, the eyes of the polychaete *Platynereis dumerilii* contain neopterin (109c), platynerepterin, and nerepterin which are probably dimers (Viscontini *et al.*, 1970). The presence of xanthopterin (109b) has been reported to be found in the ascidian *Microcosmus polymorphus* in large quantities (Karrer *et al.*, 1948; Ciereszko, 1970), but pterin (109a) and isoxanthopterin (109d) are more often found in ascidians (Gaill and Momzikoff, 1975) although usually in minute quantities.

Pterins, in addition to the aforementioned functions, may play a role in nitrogen excretion and, more interestingly, may serve as chemical communication agents. For instance, when the minnow *Phoxinus laevis* is injured, its skin releases fright substances that cause a strong fright reaction in other minnows. Such a substance has been isolated from the skin of *Phoxinus phoxinus;* it displays typical properties of pterins (Mackie, 1975; Barnett, 1977 and cited references). On the other hand, Momzikoff (1977) found xanthopterin (109b) in the skin of herrings and proposed that it could be an important factor in schooling behavior. In addition, he claimed (Momzikoff, 1977) that 7-hydroxybiopterin (109e) and 6-carboxyisoxanthopterin (109f) are present in relatively high concentration in areas where salmon assemble. He concludes that these compounds could be attracting substances. In fact, very recent research has shown that pterins may influence the behavior of some migrating fishes (Fontaine *et al.*, personal communication).

Seawater contains both flavins and pterins (Momzikoff, 1977). Among the pterins, isoxanthopterin (109d) is the most abundant with concentration ranging from 0 to 1 μg/liter. Natural populations of planktonic copepods are thought to be mainly responsible for the presence of pterins in seawater because they contain and excrete such compounds (Momzikoff, 1973; Momzikoff and Legrand, 1973) and because of the correlation between the presence of isoxanthopterin and that of copepods; but phytoplankton also contains closely related compounds (Momzikoff,

1975). Pterins in the marine environment surely play important roles both in chemical communication (see preceding paragraph) and in the complexing of trace metals (Momzikoff, personal communication).

C. 2-Aminoimidazole

Recently, Cimino *et al.*, (1974) obtained the simple 2-aminoimidazole (**110**) from the sponge *Reneira cratera*. This compound had previously been isolated from *Streptomyces eurocidicus*. In this case, addition of arginine to the growth medium considerably increased the production of **110.** This compound is thus probably derived by arginine metabolism (Seki *et al.*, 1970).

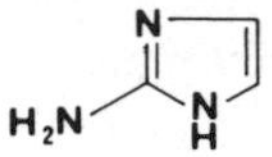

110

IX. *Cypridina* LUCIFERIN

Bioluminescence is a general phenomenon that occurs in various organisms of many phyla in marine and terrestrial environments. In both cases, at least for known examples, the process often involves a series of reactions which may be described as follows:

$$\text{Luciferin} + O_2 \xrightarrow{\text{luciferase}} \text{(peroxide)} \longrightarrow \text{(product)}^{*} + CO_2$$
$$\text{(product)}^{*} \longrightarrow \text{product} + h\nu$$

Luciferins are usually rather unstable products, and their structural determination remains a difficult problem. In the field of marine natural product chemistry, the first structure to be fully elucidated was that of *Cypridina* luciferin. As early as 1917, Harvey discovered the luciferin–luciferase reaction of *Cypridina,* an ostracod crustacean living along Japanese coasts (Goto and Kishi, 1968). Forty years later, Shimomura *et al.* (1957) isolated pure crystalline *Cypridina* luciferin for the first time. In 1966 the structure of *C.* luciferin (**111**) was determined chemically by a study of the degradation products (Kishi *et al.*, 1966a,b) and total synthesis (Kishi *et al.*, 1966c). *Cypridina* luciferin was converted to the more

$$C.\ \text{luciferin} \xrightarrow[\text{or } NH_3]{O_2 + \text{luciferase}} C.\ \text{oxyluciferin} + C.\ \text{etioluciferin} + HOOC{-}CO{-}CH(Me){-}CH_2CH_3\ (\mathbf{112})$$

$$C.\ \text{oxyluciferin} \xrightarrow{HCl} C.\ \text{etioluciferin} + HOOC{-}CO{-}CH(Me){-}CH_2CH_3$$

112

stable *C.* etioluciferin with intermediate formation of *C.* oxyluciferin. Following chemical degradation studies, structure **113** of *C.* etioluciferin was deduced and confirmed by synthesis. The formation of α-keto-β-methylvaleric acid (**112**) during hydrolysis of *C.* oxyluciferin led Kishi *et al.* (1966a) to propose structure **114** for *C.* oxyluciferin, which agreed with nmr spectral data. Goto *et al.* (1968) revised this formula slightly in **115** after comparison with synthetic analogs. The *C.* luciferin structure was deduced as **111** and confirmed by synthesis.

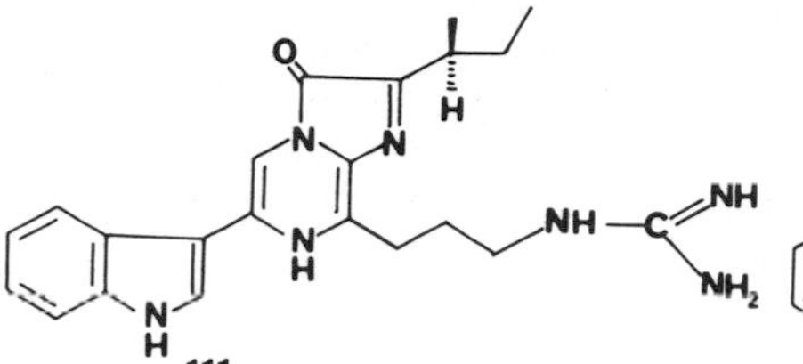

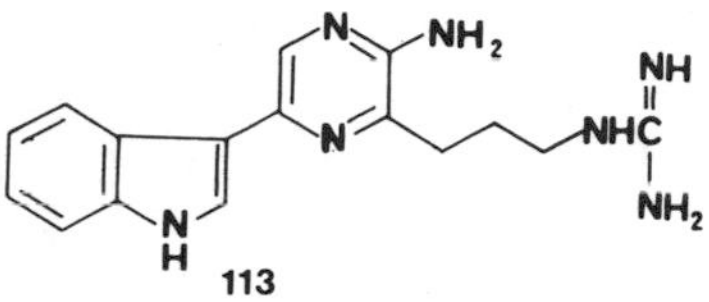

As observed by Scheuer (1973), *C.* luciferin is produced from the condensation of three amino acids (tryptophan, arginine, and isoleucine) in the manner shown in Fig. 11.

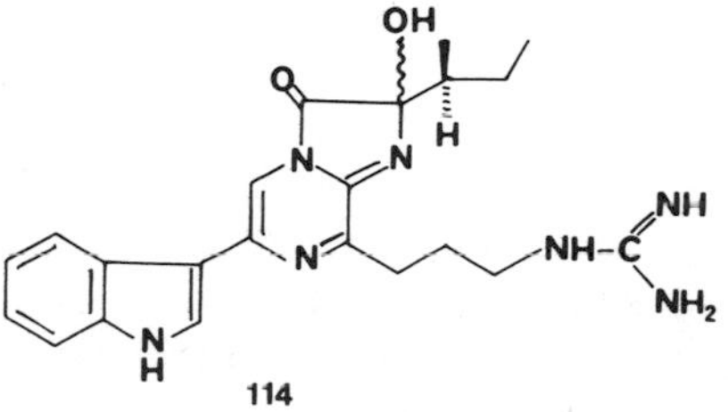

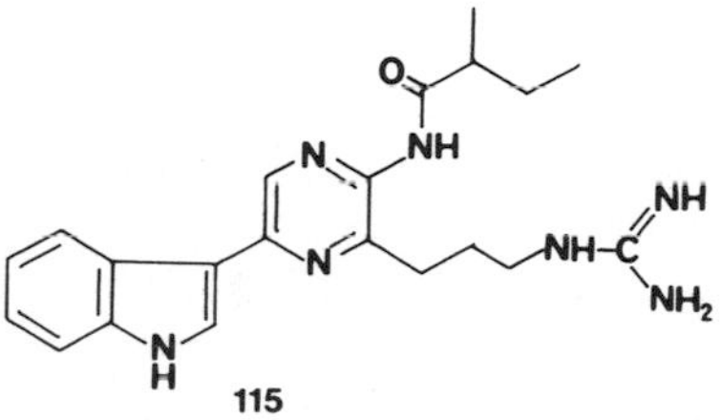

Surprisingly, other luciferins isolated from marine sources have similar structures, differing only in the amino acid substituents. The luciferin (**116**) of *Renilla reniformis* (cnidarian) contains two tyrosine and one phenylalanine residues (Inoue *et al.*, 1977a) and is identical to the decapod *Olophorus* luciferin (Inoue *et al.*, 1976a). It is also found in the fish *Neoscopelus microchir* (Inoue *et al.*, 1977b). The luciferin (**117**) of the squid *Watasenia scintillans* (Inoue *et al.*, 1976b) is also closely related.

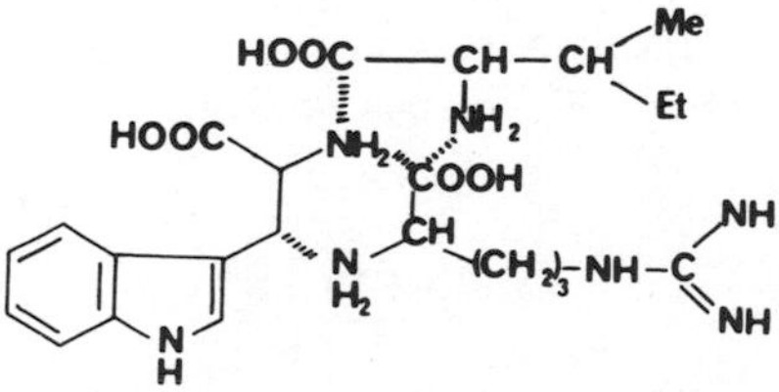

Fig. 11. Hypothetical biogenetic pathway of *Cypridina* luciferin.

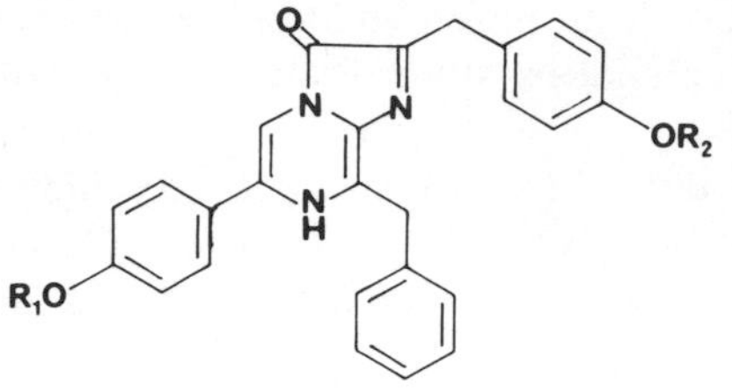

116	R₁ = H	R₂ = H
117	R₁ = SO₃H	R₂ = SO₃H

Intriguingly, the rather different *C.* luciferin and *O.* luciferin are found within the same phylum, while *O.* luciferin is also found in completely unrelated phyla. The bioluminescence of marine organisms has recently been reviewed by Goto (1979).

X. CONCLUSION

Aromatic amino acids such as tryptophan, phenylalanine, and tyrosine play an important role in the secondary metabolism of higher plants. In addition, many neurosecretory hormones (serotonin, catecholamines, etc.) are also derived from these amino acids. In the marine environment, many of the more uncommon nitrogen-containing compounds are probably derived from nonaromatic amino acids although few biogenetic studies have been performed. Arginine seems to be particularly important both as a substrate and as an amidination agent. The evolutionary significance and biological role of guanidine derivatives are subjects of considerable interest and speculation. Without doubt this is one of the richest classes of compounds that possess biological activity.

ACKNOWLEDGMENTS

The author is particularly indebted to Mr. P. Beninger for help with the original text. I also wish to thank Misses M. C. Rohan and J. Huguen for typing the manuscript and Mr. J. P. Cosson for drawing the formulas. I am grateful to Dr. Y. Robin for fruitful discussions and pertinent criticisms. Finally, I wish to thank Prof. P. Scheuer for my initiation to marine natural product chemistry and for his constant help and guidance.

REFERENCES

Ackermann, D. (1935). *Hoppe-Seyler's Z. Physiol. Chem.* **232,** 206.
Ackermann, D., and List, P. H. (1957). *Hoppe-Seyler's Z. Physiol. Chem.* **309,** 286.
Ackermann, D., and List, P. H. (1960). *Hoppe-Seyler's Z. Physiol. Chem.* **318,** 281.

Ackermann, D., and List, P. H. (1961). *Hoppe-Seyler's Z. Physiol. Chem.* **323,** 192.
Ackermann, D., and Müller, E. (1935). *Hoppe-Seyler's Z. Physiol. Chem.* **235,** 233.
Ackermann, D., and Pant, R. (1961). *Hoppe-Seyler's Z. Physiol. Chem.* **326,** 197.
Auclair, M. C., Adolphe, M., Guillou, Y., and Robin, Y. (1976). *C. R. Soc. Biol.* **170,** 65.
Baldwin, J., and Opie, A. M. (1978). *Comp. Biochem. Physiol.* **61B,** 85.
Balinski, J. B. (1970). *In* "Comparative Biochemistry of Nitrogen Metabolism" (J. W. Campbell ed.), Vol. 2, The Vertebrates, p. 577. Academic Press, New York.
Baker, J. T. (1976). *Pure Appl. Chem.* **48,** 35.
Baker, J. T., and Murphy, V. (1976). "Handbook of Marine Science; Compounds from Marine Organisms," Vol. I. CRC Press, Cleveland, Ohio.
Barnett, C. (1977). *Biosci. Commun.* **3,** 331.
Baslow, M. H. (1977). "Marine Pharmacology. A Study of Toxins and Other Biologically Active Substances of Marine Origin," p. 119. Krieger Publ., Huntington, New York.
Beatty, I. M., and Magrath, D. I. (1959). *Nature (London)* **183,** 591.
Beatty, I. M., Magrath, D. I., and Ennor, A. H. (1959). *Nature (London)* **183,** 591.
Beis, I., and Newsholme, E. A. (1975). *Biochem. J.* **152,** 23.
Bell, E. A. (1964). *Biochem. J.* **91,** 358.
Bell, E. A., and Tirimana, A. S. L. (1964). *Biochem. J.* **91,** 356.
Bergquist, P. R. (1978). "Sponges." Hutchinson Univ. Library, London.
Bergquist, P. R., and Hartman, W. D. (1969). *Mar. Biol.* **3,** 247.
Blankenship, J. E. (1976). *Perspect. Biol. Med.* **19,** 509.
Bordner, J., Thiessen, W. E., Bates, H. A., and Rapoport, H. (1975). *J. Am. Chem. Soc.* **97,** 6008.
Boyer, G. L., Schantz, E. J., and Schnoes, H. K. (1978). *J. Chem. Soc. Chem. Commun.* 889.
Boylan, D. B., and Scheuer, P. J. (1967). *Science* **155,** 52.
Braun, M., and Büchi, G. (1976). *J. Am. Chem. Soc.* **98,** 3049.
Braun, M., Büchi, G., and Bushey, D. F. (1978). *J. Am. Chem. Soc.* **100,** 4208.
Brown, G. B., Kim, Y. H., Küntzel, H., and Mosher, H. S. (1977). *Toxicon* **15,** 115.
Cariello, L., Crescenzi, S., and Prota, G. (1973). *J. Chem. Soc. Chem. Commun.* p. 99.
Cariello, L., Crescenzi, S., Prota, G., Capasso, S., Giordano, F., and Mazzarella, L. (1974a). *Tetrahedron* **30,** 3281.
Cariello, L., Crescenzi, S., Prota, G., and Zanetti, L. (1974b). *Experientia* **30,** 849.
Cariello, L., Crescenzi, S., Prota, G., and Zanetti, L. (1974c). *Tetrahedron* **30,** 3611.
Cariello, L., Crescenzi, S., Prota, G., and Zanetti, L. (1974d). *Tetrahedron* **30,** 4191.
Cariello, L., Crescenzi, S., Zanetti, L., and Prota, G. (1979). *Comp. Biochem. Physiol.* **63 B,** 77.
Carter, G. T., and Rinehart, K. L. (1978). *J. Am. Chem. Soc.* **100,** 4302.
Cheng, M. T., and Rinehart, K. L. (1978). *J. Am. Chem. Soc.* **100,** 7409.
Chevolot, L., Padua, S., Ravi, B. N., Blyth, P. C., and Scheuer, P. J. (1977). *Heterocycles* **7,** 891.
Ciereszko, L. S. (1970). *In* "Comparative Biochemistry of Nitrogen Metabolism" (J. W. Campbell, ed.), Vol. I, The Invertebrates, p. 491. Academic Press, New York.
Cimino, G., De Stefano, S., and Minale, L. (1974). *Comp. Biochem. Physiol.* **47B,** 895.
Cimino, G., De Stefano, S., Minale, L., and Sodano, G. (1975). *Comp. Biochem. Physiol.* **50B,** 279.
Corrigan, J. J. (1970). *In* "Comparative Biochemistry of Nitrogen Metabolism" (J. W. Campbell, ed.), Vol. 1, The Invertebrates, p. 387. Academic Press, New York.
Cullen, E., and Delvin, J. P. (1975). *Can. J. Chem.* **53,** 1690.
De Laubenfels, M. W. (1948). The Order Kerotosa of the Phylum Porifera. A Monographic Study. Allan Hancock Foundation Publications of the Univ. of Southern California, Occasional Paper No. 3.

Devon, T. K., and Scott, A. I. (1975). *In* "Handbook of Naturally Occuring Compounds," Vol. 1, pp. 489, 492. Academic Press, New York.

Doublet, M. O., Olomucki, A., Baici, A., and Luisi, P. L. (1975). *Eur. J. Biochem.* **59**, 185.

Fattorusso, E., Magno, S., Mayol, L., Santacroce, C., and Sica, D. (1975). *Tetrahedron* **31**, 1715.

Faulkner, D. J. (1977). *Tetrahedron* **33**, 1421.

Florkin, M., and Bricteux-Gregoire, S. (1972). *In* "Chemical Zoology" (M. Florkin and B. T. Scheer, eds.), Vol. VII, Mollusca. Academic Press, New York.

Forenza, S., Minale, L., Riccio, R., and Fattorusso, E. (1971). *J. Chem. Soc. Chem. Commun.* 1129.

Fujita, Y. (1959). *Bull. Chem. Soc. Jpn.* **32**, 439.

Fujita, Y. (1960). *Bull. Chem. Soc. Jpn.* **33**, 1379.

Fujita, Y. (1961). *J. Biochem.* (*Tokyo*) **49**, 468.

Gäde, G., and Grieshaber, M. (1975). *J. Comp. Physiol.* **102**, 149.

Gäde, G., Weeda, E., and Gabott, P. A. (1978). *J. Comp. Physiol.* **B 124**, 121.

Gaill, F., and Momzikoff, A. (1975). *Mar. Biol.* **29**, 315.

Garcia, E. E., Benjamin, L. E., and Fryer, I. R. (1973). *J. Chem. Soc. Chem. Commun.* p. 78.

Goodwin, T. W. (1971). *In* "Chemical Zoology" (M. Florkin and B. T. Scheer, eds.), Vol. VI, Arthropoda Part B, p. 289. Academic Press, New York.

Goto, T. (1979). *In* "Marine Natural Products" (P. J. Scheuer, ed.), Vol. III, pp. 180–216. Academic Press, New York.

Goto, T., and Kishi, Y. (1968). *Angew. Chem. Int. Ed.* **7**, 407.

Goto, T., Inoue, S., Sugiura, S., Nishikawa, K., Isobe, M., and Abe, Y. (1968). *Tetrahedron Lett.*, p. 4035.

Guillou, Y., and Robin, Y. (1973). *J. Biol. Chem.* **248**, 5668.

Guillou, Y., and Robin, Y. (1979). *C. R. Soc. Biol.* **173**, 576.

Haurowitz, F., and Waelsch, H. (1926). *Hoppe-Seyler's Z. Physiol. Chem.* **161**, 300.

Hille, B. (1975). *Biophys. J.* **15**, 615.

Hollenbeak, K. H., and Schmitz, F. J. (1977). *Lloydia* **40**, 479.

Inoue, S., Kakoi, H., and Goto, T. (1976a). *J. Chem. Soc. Chem. Commun.* p. 1056.

Inoue, S., Kakoi, H., and Goto, T. (1976b). *Tetrahedron Lett.* p. 2971.

Inoue, S., Kakoi, H., Murata, M., Goto, T., and Shimomura, O. (1977a). *Tetrahedron Lett.* 2685.

Inoue, S., Okada, K., Kakoi, H., and Goto, T. (1977b). *Chem. Lett.* 257.

Iseki, T. (1931). *Hoppe-Seyler's Z. Physiol. Chem.* **203**, 259.

Ito, K., and Hashimoto, Y. (1965). *Agr. Biol. Chem.* **29**, 832.

Ito, K., and Hashimoto, Y. (1966a). *Nippon Suisan Gakkaishi* **32**, 274.

Ito, K., and Hashimoto, Y. (1966b). *Nature* (*London*) **211**, 417.

Ito, K., and Hashimoto, Y. (1969). *Agr. Biol. Chem.* **33**, 237.

Ito, K., Miyazawa, K., and Hashimoto, Y. (1966). *Nippon Suisan Gakkaishi* **32**, 727.

Ito, K., Miyazawa, K., and Hashimoto, Y. (1967). *Nippon Suisan Gakkaishi* **33**, 572.

Kao, C. Y. (1972). *Fed. Proc. Fed. Am. Soc. Exp. Biol.* **31**, 1117.

Karrer, P., Manunta, C., and Schwyzer, R. (1948). *Helv. Chim. Acta* **31**, 1214.

Kazlauskas, R., Murphy, P. T., Quinn, R. J., and Wells, R. J. (1977). *Tetrahedron Lett.* p. 61.

Kishi, Y., Goto, T., Hirata, Y., Shimomura, O., and Johnson, F. H. (1966a). *Tetrahedron Lett.* p. 3427.

Kishi, Y., Goto, T., Eguchi, S., Hirata, Y., Watanabe, E., and Aoyama, T. (1966b). *Tetrahedron Lett.* p. 3437.

Kishi, Y., Goto, T., Inoue, S., Sugiura, S., and Kishimoto, H. (1966c). *Tetrahedron Lett.* p. 3445.

Kishi, Y. *et al.* (1972). *J. Am. Chem. Soc.* **94,** 9219.
Kisliuk, R. L., and Brown, J. (1979). "Chemistry and Biology of Pteridines." North-Holland Publ., Amsterdam.
Komoda, Y., Kaneko, S., Yamamoto, M., Ishikawa, M., Itai, A., and Itaka, Y. (1975). *Chem. Pharm. Bull.* **23,** 2464.
Kutscher, F., Ackermann, D., and Flössner, O. (1931). *Hoppe-Seyler's Z. Physiol. Chem.* **199,** 273.
Lederer, E. (1939). *C. R. Hebd. Seances Acad. Sci.* **209,** 528.
Lee, A. S. K., Vanstone, W. E., Markert, J. R., and Antia, N. J. (1969). *J. Fish. Res. Bd. Can.* **26,** 1185.
Liaci, L. S. (1970). *In* "Chemistry and Biology of Pteridines" (K. Iwai, M. Akino, M. Groto, and Y. Iwanami, eds.), p. 471. Int. Acad. Printing Co., Tokyo.
Mackie, A. M. (1975). *In* "Biochemical and Biophysical Perspectives in Marine Biology" (D. C. Malins and J. R. Sargent, ed.), Vol. 2, p. 82. Academic Press, New York.
Makisumi, S. (1961). *J. Biochem. (Tokyo)* **49,** 284.
Matsumoto, M., Fujiwara, M., Mori, A., and Robin, Y. (1977). *C. R. Soc. Biol.* **171,** 1226.
Momzikoff, A. (1973). *Cahiers Biol. Mar.* **14,** 323.
Momzikoff, A. (1975). *In* "Chemistry and Biology of Pteridines" (W. Pfleiderer, ed.), p. 871. de Gruyter, Berlin and New York.
Momzikoff, A. (1977). These, Doctorat d'Etat. Univ. Pierre et Marie Curie. Paris 6.
Momzikoff, A., and Legrand, J. M. (1973). *Cahiers Biol. Mar.* **14,** 249.
Moore, E., and Wilson, D. W. (1937). *J. Biol. Chem.* **119,** 573.
Moore, R. E. (1977). *Bioscience* **27,** 797.
Morizawa, K. (1927). *Acta. Schol. Med. Univ. Imp. Kioto* **9,** 285 [*Chem. Abstr.* **22,** 3705 (1928).]
Mosher, H. S., Fuhrman, F. A., Buchwald, H. D., and Fisher, H. G. (1964). *Science* **144,** 1100.
Narahashi, T. (1972). *Fed. Proc. Fed. Am. Soc. Exp. Biol.* **31,** 1124.
Needham, A. E. (1970). *In* "Comparative Biochemistry of Nitrogen Metabomism" (J. W. Campbell, ed.), Vol. 1, The Invertebrates, p. 226. Academic Press, New York.
Noguchi, T., and Hashimoto (1973). *Toxicon* **11,** 305.
Noguchi, T., Konosu, S., and Hashimoto, Y. (1969). *Toxicon* **7,** 325.
Pavelka, L. A., Kim, Y. H., and Mosher, H. S. (1977). *Toxicon* **15,** 135.
Pelter, A., Ballantine, J. A., Ferrito, V., Jaccarini, V., Psaila, A. F. and Schembri, P. J. (1976). *J. Chem. Soc. Chem. Commun.* p. 999.
Pfleiderer, W. (1964). *Angew. Chem. Int. Ed. Engl.* **3,** 114.
Pfleiderer, W. (1975). "Chemistry and Biology of Pteridines." De Gruyter, Berlin, New York.
Potier, P. (1978). *Rev. Latino-Am. Quim* **9,** 47.
Prota, G. (1979). *In* "Marine Natural Products" (P. J. Scheuer, ed.), Vol. III, pp. 141–174. Academic Press, New York.
Quadrifoglio, F., Crescenzi, V., Prota, G., Cariello, L., Di Marco, A., and Zunino, F. (1975). *Chem. Biol. Interact.* **11,** 91.
Ritchie, J. M. (1978). *In* "Cell Membrane Receptors for Drugs and Hormones: Multidisciplinary Approach" (R. W. Straub and L. Bolis, eds.), p. 227. Raven, New York.
Robin, Y. (1964a). *Comp. Biochem. Physiol.* **12,** 347.
Robin, Y. (1964b). *Biochim. Biophys. Acta* **93,** 206.
Robin, Y. (1964c). *C. R. Soc. Biol.* **158,** 490.
Robin, Y. (1974). *Biosystems* **6,** 49.
Robin, Y. (1980). *In* "Actualités de Biochimie marine" (Y. Le Gall, ed.), Vol. 2, pp. 255–270. Colloque GABIM-1977-BREST, Edition du CNRS, Paris.
Robin, Y., and Guillou, Y. (1977). *Anal. Biochem.* **83,** 45.

Robin, Y., and Guillou, Y. (1980). "Oceanis," Vol. 5, Fascicule hors-série. Actes du colloque ATP Oceanographie chimique, p. 575.
Robin, Y., and Roche, J. (1965). *Comp. Biochim. Physiol.* **14**, 453.
Robin, Y., and Thoai, N. V. (1961a). *C. R. Hebd. Seances Acad. Sci.* **252**, 1224.
Robin, Y., and Thoai, N. V. (1961b). *Biochim. Biophys. Acta* **52**, 233.
Robin, Y., and Thoai, N. V. (1962). *Biochim. Biophys. Acta* **63**, 481.
Robin, Y., Thoai, N. V., Pradel, L. A., and Roche, J. (1956). *C. R. Soc. Biol.* **150**, 1892.
Robin, Y., Thoai, N. V., and Roche, J. (1957). *C. R. Soc. Biol.* **151**, 2015.
Roche, J., and Robin, Y. (1954). *C. R. Soc. Biol.* **148**, 1541.
Roche, J., Audit, C., and Robin, Y. (1965). *C. R. Hebd. Seances Acad. Sci.* **260**, 7023.
Rose, A. F. (1975). PhD. Dissertation, Univ. of Hawaii.
Rose, A. F., Scheuer, P. J., Springer, J. P., and Clardy, J. (1978). *J. Am. Chem. Soc.* **100**, 7665.
Sangster, A. W., Thomas, S. E., and Tingling, N. L. (1975). *Tetrahedron* **31**, 1135.
Sato, M., Sato, Y., and Tsuchiya, Y. (1977) *Nippon Suisan Gakkaishi* **43**, 1077.
Sato, M., Sato, Y., and Tsuchiya, Y. (1978). *Nippon Suisan Gakkaishi* **44**, 247.
Schantz, E. J. *et al.* (1975). *J. Am. Chem. Soc.* **97**, 1238.
Scheuer, P. J. (1973). "Chemistry of Marine Natural Products." Academic Press, New York.
Scheuer, P. J. (1975). Lloydia **38**, 1.
Scheuer, P. J. (1977). *Accounts Chem. Res.* **10**, 33.
Scheumack, D., Howden, M. E. H., Spence, I., and Quinn, R. J. (1978). *Science* **199**, 188.
Schwartz, R. E., Yunker, M. B., and Scheuer, P. J. (1978). *Tetrahedron Lett.* 2235.
Schwartz, R. E., Yunker, M. B., Scheuer, P. J., and Ottersen, T. (1979). *Can. J. Chem.* **57**, 1707.
Seki, Y., Nakamura, T., and Okami, Y. (1970). *J. Biochem. (Tokyo)* **67**, 389.
Sharma, G. M., and Burkholder, P. R. (1971). *J. Chem. Soc. Chem. Commun.* 151.
Sharma, G., and Magdoff-Fairchild (1977). *J. Org. Chem.* **42**, 4118.
Sharma, G. M., Vig, B., and Burkholder, P. R. (1970). *Proc. Conf. Food-Drugs from the Sea* p. 307. Marine Technology Society, Washington, D. C.
Shimizu, Y. (1978). *In* "Marine Natural Products." (P. J. Scheuer, *ed.*), Vol. 1, pp. 1–42.
Shimizu, Y., Hsu, C. P., Fallon, W. E., Oshima, Y., Miura, I., and Nakanishi, K. (1978b). *J. Am. Chem. Soc.* **100**, 6791.
Shimizu, Y., Buckley, L. J., Alam, M., Oshima, Y. Fallon, W. E., Kasai, H., Miura, I., Gullo, V. T. and Nakanishi, K. (1976). *J. Am. Chem. Soc.* **98**, 5414.
Shimomura, O., Goto, T., and Hirata, Y. (1957). *Bull. Chem. Soc. Jpn.* **30**, 929.
Stempien, M. F., Nigrelli, R. F., and Chib, J. S. (1972). 164th ACS Meeting, Abstracts, MEDI 21.
Suzuki, T., and Muraoka, S. J. (1954). *J. Pharm. Soc. Jpn.* **74**, 171.
Tanino, H., Inoue, S., Aratani, M., and Kishi, Y. (1974). *Tetrahedron Lett.* p. 335.
Tanino, H., Nakata, T., Kaneko, T., and Kishi, Y. (1977). *J. Am. Chem. Soc.* **99**, 2818.
Taylor, K. M., Davis, P. A., Baird-Lambert, J. A., Murphy, P. T., and Wells, R. J. (1978). Roche Research Institute of Marine Pharmacology, Dee Why, New South Wales, Australia, 1st Research Report, p. 8.
Thoai, N. V. (1965). *In* "Comprehensive Biochemistry" (M. Florkin and E. H. Stotz, eds.), Vol. 6. pp. 208–253. Elsevier, Amsterdam.
Thoai, N. V., and Robin, Y. (1954a). *Biochim. Biophys. Acta* **13**, 533.
Thoai, N. V., and Robin, Y. (1954b). *Biochim. Biophys. Acta* **14**, 76.
Thoai, N. V., and Robin, Y. (1959a). *Biochim. Biophys. Acta* **35**, 446.
Thoai, N. V., and Robin, Y. (1959b). *Bull. Soc. Chim. Biol.* **41**, 735.

Thoai, N. V., and Robin, Y. (1961). *Biochim. Biophys. Acta* **52,** 221.
Thoai, N. V., and Robin, Y. (1969). *In* "Chemical Zoology" (M. Florkin and B. T. Scheer, eds.), Vol. 4, pp. 163–203. Academic Press, New York.
Thoai, N. V., and Roche, J. (1964). *Biol. Rev.* **39,** 214.
Thoai, N. V., Roche, J., and Robin, Y. (1953a). *Biochim. Biophys. Acta* **11,** 403.
Thoai, N. V., Roche, J., Robin, Y., and Thiem, N. V. (1953b). *Biochim. Biophys. Acta* **11,** 593.
Thoai, N. V., Zappacosta, S., and Robin, Y. (1963a). *Comp. Biochem. Physiol.* **10,** 209.
Thoai, N. V., Di Jeso, F., and Robin, Y. (1963b). *C. R. Hebd. Seances Acad. Sci. Ser.* **256,** 4525.
Thoai, N. V., Regnouf, F., and Olomucki, A. (1967). *Bull. Soc. Chim. Biol.* **49,** 805.
Thoai, N. V., Huc, C., Pho, D. B., and Olomucki, A. (1969). *Biochim. Biophys. Acta* **191,** 46.
Thoai, N. V., Robin, Y., and Guillou, Y. (1972). *Biochemistry* **11,** 3890.
Ulbricht, W. (1977). *Electron. Phenom. Biol. Memb. Level, Proc. Int. Meeting Soc. Chim. Phys., 29th, 1976* (E. Roux, ed.), p. 203. Elsevier, Amsterdam.
Von Lendenfeld, R. (1889). "A Monograph of the Horny Sponges." Trübner, London.
Viscontini, M., Hummel, W., and Fisher, A. (1970). *Helv. Chim. Acta* **53,** 1207.
Watts, D. C. (1977). GABIM Meeting, Brest.
Ziegler, I., and Harmsen, R. (1969). *J. Insect Physiol.* **6,** 139.

Chapter 3

Phenolic Substances

TATSUO HIGA

I. INTRODUCTION

Phenolic substances are widely distributed in nature, particularly in the plant kingdom. Their marine occurrence has been reported from a number of plants and animals.* Among marine plants, brown and red algae have

*The phenolic compounds discussed in this chapter, except bacterial metabolites, are compiled in Tables A1 (plants) and A2 (animals).

93

MARINE NATURAL PRODUCTS
Copyright © 1981 by Academic Press, Inc.
All rights of reproduction in any form reserved.
ISBN 0-12-624004-3

provided a host of phenolic and other secondary metabolites, which is a reflection of the fact that these plants have been favorite research targets of marine natural products chemists in recent years. Relatively few species of green and blue-green algae and marine phanerogames have been chemically explored, and that circumstance also is reflected in the fact that few phenolic substances have been isolated from these sources. Among marine animals, sponges and echinoderms exceed other animal phyla in the reported number of phenolic metabolites. Other major sources of phenolic compounds include mollusks, hemichordates, annelids, and coelenterates; except for coelenterates, not many species of these animals have been chemically investigated.

Some 170 phenols and phenolic ethers are covered in this chapter. Although by no means comprehensive, Table 1 outlines the current status

TABLE 1

Marine Organisms and Structural Classes of Phenolic Substances[a]

Organisms	Number of phenolic substances isolated	Structural classes
Plants		
Red algae	37	Simple bromophenols, phenolic sesquiterpenes
Brown algae	42	Phloroglucinol oligomers, polyprenyl hydroquinones
Green algae	8	Brominated geranylhydroquinones
Blue-green algae	9	Simple bromophenols, complex phenols (e.g., aplysiatoxins)
Sea grasses	7	Phenylpropanoids, flavonoids
Animals		
Sponges	35	Dibromotyrosine-derived phenols, polyprenyl hydroquinones
Echinoderms	(38)[b]	Naphtho- and anthraquinones, naphthopyrones
Mollusks	15	Phenolic sesquiterpenoids, complex phenols
Hemichordates	11	Simple bromophenols and oligomers
Annelids	9[c]	Simple bromophenols and dimers
Pholonids	2	Simple bromophenols
Coelenterates	9	Phenolic sesquiterpenoids
Protochordates	1	Geranylhydroquinone
Bacteria	3	Simple phenols

[a] Phenolic luciferins are not included.
[b] Not discussed in this review.
[c] The number includes hallachrome and arenicochrome (see Scheuer, 1973).

of phenolic compounds recorded from marine organisms. Structural classes listed in Table 1 are meant to be representative only. For individual compound-species relationships the reader is referred to Tables A1 and A2 in the Appendix. In order to compile fairly comprehensive tables for phyletic distribution of phenolic substances, an attempt has been made to collect as many of these compounds as possible, but some were deliberately omitted. These are naphthoquinones, naphthopyrones, and anthraquinones, all of which are metabolites of a single phylum, Echinodermata. Since the time of previous reviews (Scheuer, 1973; Thomson, 1971; Grossert, 1972), virtually no progress has been reported on these classes of compounds. Also omitted are phenolic luciferins, the light emitters of bioluminescent organisms, which were reviewed by Goto in Volume III of this series. Other phenols which appeared in Volumes I and II are briefly treated with an indication of their occurrence. Altogether, over 220 phenolic substances have been described from marine organisms to date. Structurally, these compounds vary from simple C_6 phenols to those linked to complex skeletons, such as the aplysiatoxins and the polycyclic polyprenyl hydroquinones, and from monohydric to polyhydric phenols, such as phloroglucinol oligomers. Most of them are structurally rather simple, and structure elucidation presents no difficulty. Since structural determination of some complex phenols has been treated in preceding volumes, only minimal discussion in this chapter is devoted to structural elucidation. Instead, the distribution and structural relation among the phenols are emphasized.

Perhaps the most obvious characteristic of marine as compared with terrestrial phenols is the presence of an abundance of simple bromophenols in both the marine fauna and flora. Many of these, as well as phenols in general, are biologically active. Some bromophenols obtained from red algae are known to inhibit algal growth above certain concentration levels, while at lower concentration growth-stimulating effects have been observed (Fries, 1973; McLachlan and Craigie, 1966). Antimicrobial activity is the most common biological property observed for phenolic substances. Recent reviews on antibiotics from marine organisms list a host of phenolic substances (Faulkner, 1978; Glombitza, 1979). Other interesting activities of phenols include those of anticancer agents (Chang and Weinheimer, 1977; Schmitz et al., 1977; Mynderse et al., 1977; Mynderse and Moore, 1978), larval settling inducers (Kato et al., 1975b), and an alarm pheromone (Sleeper and Fenical, 1977). Of current interest is the ecological significance of phenols in the marine environment, which has been speculated upon, but has not been fully investigated (Fenical, 1975; Sieburth and Jensen, 1969; Craigie and MaLachlan, 1964).

II. SHIKIMIC ACID-DERIVED SIMPLE PHENOLS

Hydroxylation patterns of natural phenolic compounds reveal, in a number of instances, their biosynthetic origin (for biosynthesis, see Geissman, 1967; Geissman and Crout, 1969). Many phenols are derived by a shikimic acid pathway via phenylalanine or tyrosine. Monohydric phenols derived from these amino acids are usually *p*-hydroxy compounds. In polyhydric aromatic compounds of shikimic acid origin the hydroxylation patterns are typically those of catechol (1), hydroquinone (3), and pyrogallol (4), while acetate-derived phenols have characteristically meta-disposed hydroxyls, as in resorcinol (2) and phloroglucinol (5).

In this section simple phenols which appear to be derived by the shikimic acid pathway have been discussed. Some seemingly complex representatives, which, however, contain a phenolic skeleton obviously of this class, have been included.

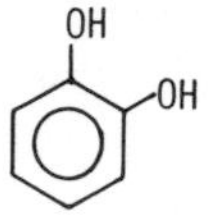

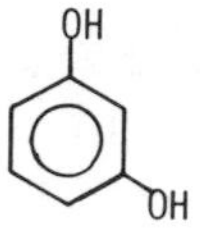

A. C$_6$ Phenols

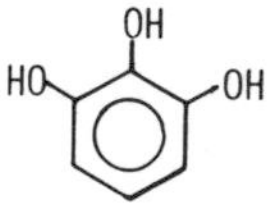

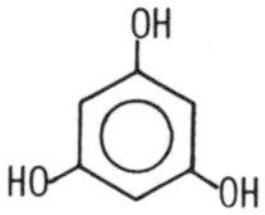

Three monohydric bromophenols are known from marine invertebrates. Ashworth and Cormier (1967) isolated 2,6-dibromophenol (6; mp 52°) from the enteropneust *Balanoglossus biminiensis* collected at Sapelo Island, Georgia. The phenol 6 was shown to be responsible for the characteristic odor that is a conmon feature of many species belonging to the class Enteropneusta (acorn worms) and said to be reminiscent of iodoform. Significantly, an individual animal, 25–30 cm long, was estimated to contain 10–15 mg of 6. The phenol 6 has also been identified from other species of the same genus, *Balanoglossus carnosus* and *Balanoglossus misakiensis*, found in Kyushu, Japan (Higa *et al.*, 1980). From *B. carnosus* we isolated another odoriferous compound, 2,4-dibromophenol (7; mp 40°) which in fact has a stronger odor than does 6. In our previous study of the odoriferous principle of the Hawaiian enteropneust *Ptycho-*

dera flava laysanica, we detected neither **6** nor **7.** The simplest bromophenol isolated from that species was 2,4,6-tribromophenol (**8**; mp 91°–94°) which, however, was odorless (Higa and Scheuer, 1977). Subsequent work revealed that the odor of *P. flava* was due to 3-haloindoles (Higa and Scheuer, 1975b). Our comparative studies of several species indicated that the iodoform-like odor of acorn worms is due to bromophenols in the genus *Balanoglossus* and to 3-haloindoles in the genera *Ptychodera* and *Glossobalanus* (Higa *et al.,* 1980).

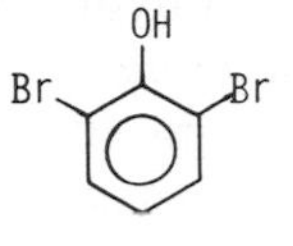

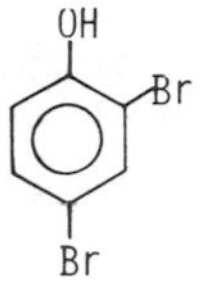

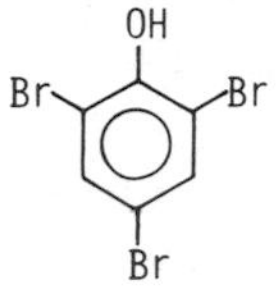

Bromophenols **6** and **8** have also been identified as antiseptic constituents of the phoronid worm *Phoronopsis viridis* (Sheikh and Djerassi, 1975). Furthermore, compound **8** has been isolated from *B. carnosus* and *Glossobalanus* sp. (Higa *et al.,* 1980) and detected by gas chromatography–mass spectroscopy (GC–MS) from an extract of the polychaete *Lanice conchilega* (Weber and Ernst, 1978).

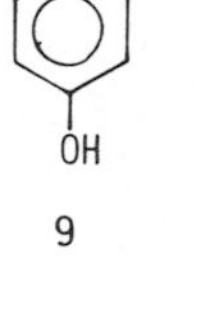

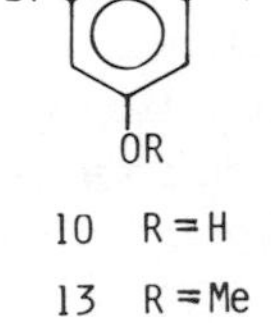

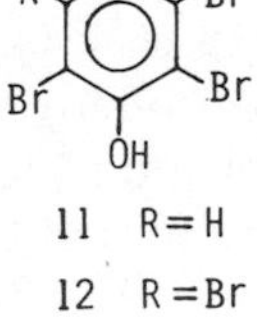

A C_6 dihydric phenol hydroquinone (**3**) and its bromo derivatives have been isolated from four species of enteropneusts: hydroquinone (**3**), bromohydroquinone (**9**; mp 114°–114.5°), and 2,6-dibromohydroquinone (**10**; mp 162°–163°) from *Glossobalanus* sp.; tribromohydroquinone (**11**; mp 135°–137°) from *B. carnosus;* **10** from *B. misakiensis;* and **11** and tetrabromohydroquinone [**12**; mp 254°–255° (dec)] from *P. flava* (Higa *et al.,* 1980). Hydroquinone **10** was distinguished from its 2,5-dibromo isomer by the nmr data of the dimethyl derivative **13,** which showed two distinct methoxy signals. The structures of **9, 11,** and **12** were confirmed by synthesis from **3.** The biogenesis of these phenols (**6–12**) in acorn worms has been proposed by a scheme involving tyrosine as the common precursor, in which several bromination and/or oxidative cleavage steps lead to all observed products (Higa *et al.,* 1980). Sheikh and Djerassi (1975) suggested a mechanism involving peroxidase-catalyzed bromination of p-hydroxybenzoic acid and subsequent decarboxylation for the biogenesis of **6** and **8** in *Phoronopsis viridis*.

One of the most uncommon phenolic metabolites is 2-methoxy-4,6-dinitrophenol (**14**; mp 124°–125°) isolated as one of the antimicrobial constituents of the red alga *Marginisporum aberrans* (Ohta and Takagi, 1977b). The compound was characterized by spectroscopic data and confirmed by comparison with an authentic sample. It appears to be the first reported marine phenol containing nitro groups.

Cimino *et al.* (1974) have isolated underivatized hydroxyhydroquinone (**15**; mp 138°–140°) from the sponge *Axinella polypoides*. The compound was identified by comparison with authentic material. It was the first example of the occurrence of hydroxyhydroquinone free in nature.

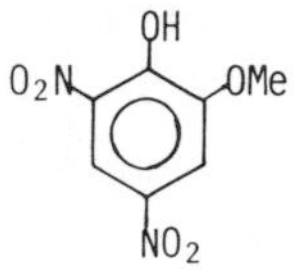

14 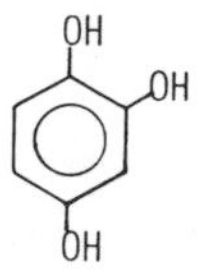   15

B. C$_7$ Phenols

The class of C$_7$ phenols includes cresols and hydroxylated benzyl alcohols, benzaldehydes, and benzoic acids. Examples of these different oxidation states have been found in the marine fauna and flora. They are all mono- and dihydric phenols with free phenolic hydroxy groups, and many of them are brominated phenols. All monohydric phenols are p-hydroxy derivatives, while the dihydric phenols are brominated 3,4-dihydroxy derivatives.

1. Monohydric Phenols

Detection of p-cresol (**16**) by GC has been reported from several species of green, brown, and red algae (Katayama, 1961), but it has never been isolated. Recently, its dibromo derivative (**17**) was detected by GC–MS in an extract of the polychaete *Lanice conchilega* (Weber and Ernst, 1978).

Fenical and McConnell (1976) isolated p-hydroxybenzyl alcohol (**18**; mp 115°–116°) and p-hydroxybenzaldehyde (**19**; mp 114°–116°) from the methylene chloride extract of the red alga *Dasya pedicellata* var. *stanfordiana*. Aldehyde **19** was responsible for the antibiotic activity of the algal extract. The same aldehyde (**19**) was also identified as an active constituent of another red alga, *Marginisporum aberrans* (Ohta and Takagi, 1977b), and of a culture of *Chromobacterium* sp. (Andersen *et al.*, 1974). Brominated derivatives of **18** are common in many species of the red algal

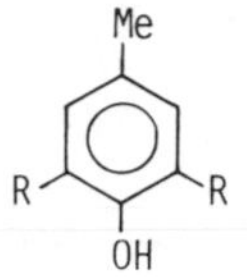

Me structures with R substituents and OH:

16 R = H
17 R = Br

R substituent structure with OH:

18 R = CH$_2$OH
19 R = CHO

family Rhodomelaceae. Thus, 3-bromo-4-hydroxybenzyl alcohol (**20**) and 3,5-dibromo-4-hydroxybenzyl alcohol (**21**) have been detected in several species of red algae (Glombitza *et al.*, 1974). Pedersén *et al.* (1974) also detected **21** by GC–MS in the extracts of four species of Rhodomelaceae. Bromophenol **21** (mp 115°–116°) was first isolated by Craigie and Gruenig (1967) from red algae *Odonthalia dentata* and *Rhodomela confervoides* and was assigned its structure mainly on the basis of the nmr spectrum of the diacetate of **21**. In other than red algae, bromophenol **21** has been detected in the brown alga *Fucus vesiculosus* (Pedersén and Fries, 1975) and in a culture of the blue-green alga *Calothrix brevissima* (Pedersén and Dasilva, 1973). Also detected in the culture were 2,3,5-tribromo-4-hydroxybenzyl alcohol (**22**) and 2,3,5-tribromo-4-hydroxybenzaldehyde (**23**). These bromophenols are not exclusive algal metabolites. We have identified 3,5-dibromo-4-hydroxybenzyl alcohol (**21**) and 3,5-dibromo-4-hydroxybenzaldehyde (**24**; mp 182°–186°) in the chloroform extracts of the polychaete *Thelepus setosus* (Higa and Scheuer, 1974, 1975a).

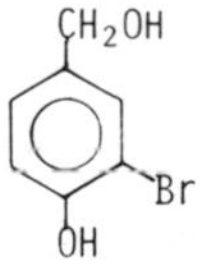

CH$_2$OH structure with Br and OH:

20

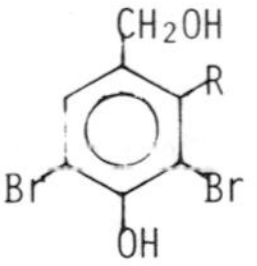

CH$_2$OH structure with Br, R, Br and OH:

21 R = H
22 R = Br

CHO structure with Br, R, Br and OH:

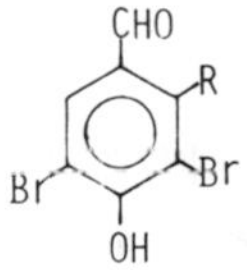

23 R = Br
24 R = H

Methyl ether **25** (mp 70°–71°) of **21**, which has been detected by Pedersén *et al.* (1974) and later isolated by Kurata *et al.* (1976), is presumably an artifact formed by methylation of **21** with methanol used in those extractions. As will also be shown by other examples, such ethers of benzyl alcohols are often artifacts of isolation, especially when an alcoholic solvent is used under acidic conditions. Nevertheless, the debromo analog of **25**, *p*-(methoxymethyl)phenol (**26**), has been found responsible for the foul odor of the red alga *Martensia fragilis* (Moore, 1977).

In contrast to numerous studies on marine algae, sea grasses or marine phanerogames have received scant attention from natural products

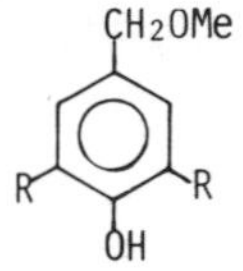

CH$_2$OMe structure with R groups and OH:

25   R = Br
26   R = H

chemists. Sea grasses are higher plants and are therefore expected to exhibit some chemical features that are present in terrestrial plants. Recent work by Cariello *et al.* (1979) shows the existence of such chemical links between the higher plants of land and sea. These authors isolated several phenolic compounds from two species of marine phanerogames. Most of them are phenylpropanoids known from terrestrial plants and will be discussed in Section II,D and E. Three simple phenols to be considered here are benzoic acid derivatives isolated from the phanerogame *Posidonia oceanica.* They are *p*-hydroxybenzoic acid (27), vanillic acid (28), and the complex phenol 29, glucosyl 4-(4-hydroxybenzoxy)-3-methoxybenzoate. The structure of the latter was determined by spectral analysis and by acid hydrolysis of the peracetate of 29, which yielded 27, 28, and glucose.

Another benzoic acid derivative, *n*-propyl 4-hydroxybenzoate (30; mp 96°), has been identified in the culture of *Chromobacterium* sp. (Andersen *et al.*, 1974).

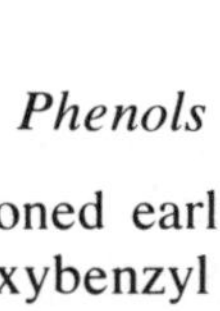

Structures for compounds:

27   R = H
28   R = OMe
29 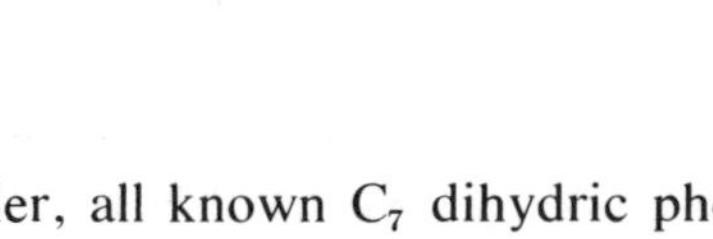
30 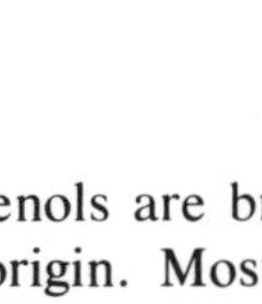

2. Dihydric Phenols

As mentioned earlier, all known C$_7$ dihydric phenols are brominated 3,4-dihydroxybenzyl alcohol derivatives of algal origin. Most of these phenols (31–40) had been discovered by early 1970, and virtually no new compounds have been reported since the reviews by Scheuer (1973) and by Fenical (1975). These bromophenols have been detected in several families of red algae, but they are most common in the members of the family Rhodomelaceae (Pedersén *et al.*, 1974). In one case, 2,3-dibromo-4,5-dihydroxybenzyl alcohol (lanosol, 33) was found in the brown alga *Fucus vesiculosus* (Pedersén and Fries, 1975). Since Saito and Ando

(1955) first identified 3-bromo-4,5-dihydroxybenzaldehyde (**31**; mp 230°) in *Polysiphonia morrowii*, it has been detected in several species of the genus *Polysiphonia* (Stoffelen *et al.*, 1972; Glombitza *et al.*, 1974; Pedersén *et al.*, 1974; Kurata *et al.*, 1976). More common than **31** is 2,3-dibromo-4,5-dihydroxybenzaldehyde (**32**; mp 203°–205°) which was first isolated by Katsui *et al.*, (1967) and since then has been detected in five genera of red algae encompassing 15 species (Stoffelen *et al.*, 1972; Pedersén *et al.*, 1974; Glombitza *et al.*, 1974; Kurata and Amiya, 1975, 1977; Lundgren *et al.*, 1979). Lanosol (**33;** mp 129°–130°) is one of the most widely distributed bromophenols in red algae. It has been isolated from *Odonthalia dentata* and *Rhodomela confervoides* (Craigie and Gruenig, 1967), *O. corymbifera* (Kurata *et al.*, 1973), *Rytiphlea tinctoria* (Chevolot-Magueur *et al.*, 1976), *Rhodomela larix* (Kurata and Amiya, 1977), and *Polysiphonia brodiaei* (Lundgren *et al.*, 1979). Furthermore, **33** has been detected in 13 other species (Pedersén *et al.*, 1974; Glombitza *et al.*, 1974). Glombitza's group (1974) has also identified by GC–MS 2,3,6-tribromo-4,5-dihydroxybenzyl alcohol (**34**) in alcoholic extracts of *P. lanosa* and *Rhodomela subfusca*, while both Pedersén's (1974) and Glombitza's groups (1974) have found 3-bromo-4,5-dihydroxybenzyl alcohol (**35**) in several species of red algae.

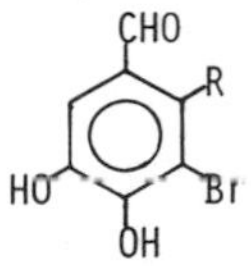

31 R = H
32 R = Br

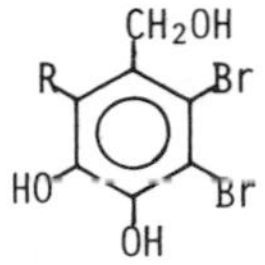

33 R = H
34 R = Br

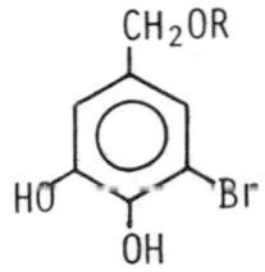

35 R = H
36 R = Me

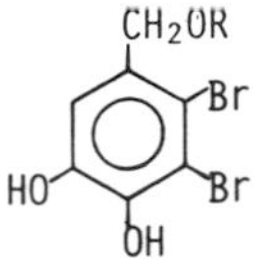

37 R = Me
38 R = Et
39 R = n-Pr

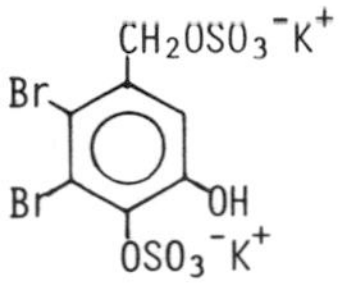

40

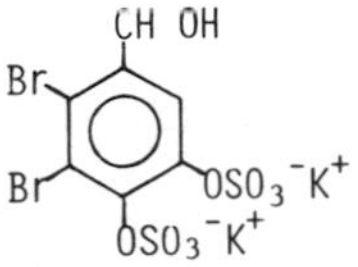

41

Benzyl methyl ethers (**36–39**), which have been found by many workers, are now believed to be artifacts. 3-Bromo-4,5-dihydroxybenzyl methyl ether (**36**) and 2,3-dibromo-4,5-dihydroxybenzyl methyl ether (**37**) have been found only when methanol was used for extraction (Kurata *et al.*, 1973, 1976; Katsui *et al.*, 1967), while the corresponding ethyl ether **38** has been detected in ethanolic extraction (Chevolot-Magueur, 1976; Pedersén and Fries, 1974). Scheuer has already pointed out that **37,** which had been isolated by Katsui *et al.* (1967), might be an artifact. Weinstein *et al.* (1975) have demonstrated that **37** and **33** could be readily formed by brief treatment of an algal metabolite dipotassium 2,3-dibromo-5-hydroxy-benzyl-1′,4-disulfate **(40)** with methanol and water, respectively. The authors suggested that even 2,3-dibromo-4,5-dihydroxybenzaldehyde **(32)** might also be formed by oxidation of **33** during isolation.

The structure of a phenolic sulfate to which Hodgkin *et al.* (1966) had assigned structure **41** has been revised to **40** by Glombitza and Stoffelen (1972). The disulfate was isolated from *Polysiphonia lanosa*—the same alga from which the Canadian group obtained the same sulfate—and treated with diazomethane to form **42,** which in turn was hydrolyzed under acidic or enzymatic conditions using an arylsulfatase to yield 2,3-di-bromo-4-hydroxy-5-methoxybenzyl alcohol **(43).** (Scheme 1). The latter product was identified by comparison with an authentic sample.

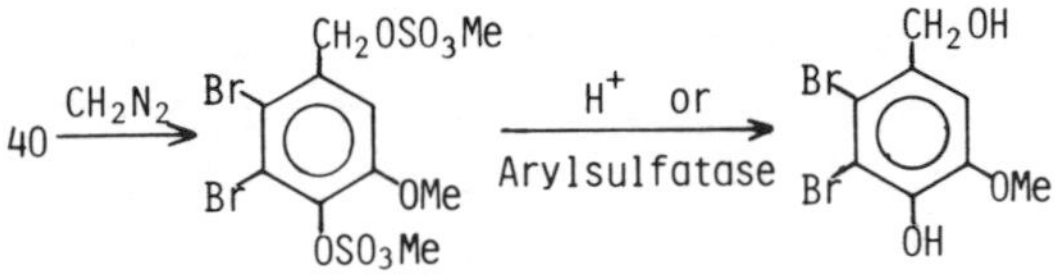

Scheme 1

From other than *P. lanosa,* dipotassium 2,3-dibromo-5-hydroxybenzyl-1′,4-disulfate **(40)** has been isolated from *Odonthalia corymbifera* (Kurata *et al.,* 1973), *Rhodomela larix* (Weinstein *et al.,* 1975), and *R. subfusca* (Kurata and Amiya, 1975). Yields of **40** are usually very high (9.2% dry weight of *R. subfusca*). Lanosol **(33)** seems to occur mostly in the form of the disulfate **40** in many species of red algae. Lanosol is highly toxic to bacteria and algae (McLachlan and Craigie, 1966), but disulfate **40** is inactive against these organisms (Glombitza and Stoffelen, 1972). This difference appears to have significant ecological implications; that is, algae presumably store the phenolic metabolites in nontoxic form as in **40** and exude it in the form of lanosol or **40,** which could be hydrolyzed to lanosol in surrounding water. Biosynthesis of these bromophenols from L-tyrosine has been proposed based on the incubation study of labeled L-tyrosine in cell-free homogenates of *Odonthalia floccosa* (Manley and Chapman, 1978).

C. C$_8$ Phenols

Many β-phenylethylamine-related alkaloids such as tyramine are known from terrestrial organisms. Some of them (e.g., epinephrine) exhibit important physiological activities. Relatively few such phenolic amines have been reported from marine sources. Tocher and Craigie (1966) isolated 3-hydroxytyramine (dopamine, **44**) from the green alga *Monostroma fuscum* and established its identity by comparison of physical and chemical properties with those of an authentic sample. 3-Hydroxytyramine (**44**) has been reported as the principal phenol and the natural substrate for phenolase in the alga. Hordenine (**45**; mp 117°) has been obtained by acidic methanol extraction of the red alga *Phyllophora nervosa* (Güven *et al.*, 1970).

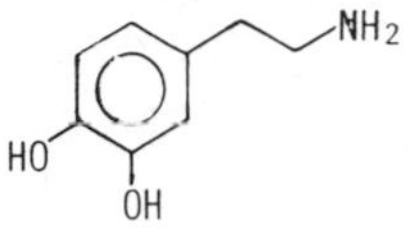

44 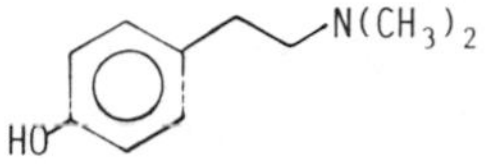   45

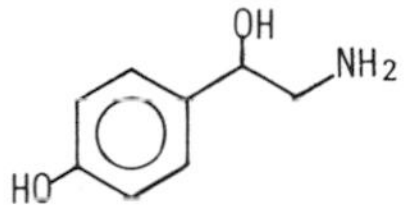

46 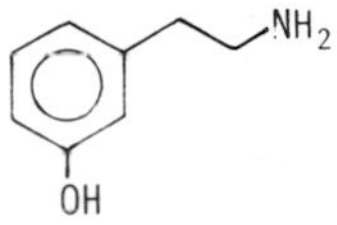   47

Erspamer (1952) identified octopamine (**46**; mp 117°–179°) in the extracts of posterior salivary glands of *Octopus vulgaris*. Compound **46** has been shown to have the same configuration as D-(−)-norepinephrine. *m*-Tyramine (**47**) has also been identified in an extract of *Palythoa* sp. (Sheikh, 1969). A novel octopamine derivative containing 2-oxazolidone rings has been found in the sponge *Verongia lacunosa*. Borders *et al.* (1974) isolated a compound designated LL-PAA216, 5-[3,5-dibromo-4-[(2-oxo-5-oxazolidinyl)methoxyphenyl]-2-oxazolidinone (**48**, mp 222°–225°, $[\alpha]_D$ +8.9°), from an extract obtained by steeping the sponge in 75% ethanol for several weeks at room temperature. They determined its structure by spectral data of **48** and of its sole hydrolysis product; resulting from vigorous acid treatment. Compound **48** had no significant antimicrobial activity.

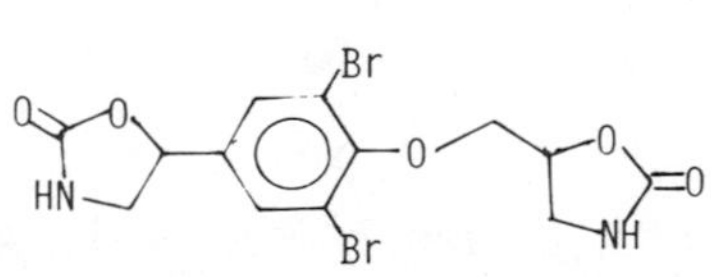

48

Members of the sponge family Clionidae have the ability to burrow into biogenic calcium carbonate. The mechanism of this excavation is believed to involve chemical dissolution (Goreau and Hartman, 1963). Recently, Andersen (1978) isolated from the burrowing sponge *Cliona celata* a trihydroxystyrylamine derivative, clionamide (**49**; $[\alpha]_D$ +32.1°), which may function as a burrowing agent by chelating with calcium ion. Andersen (1978) initially isolated tetraacetyl clionamide (**50**; mp 209°–211°, $[\alpha]_D$ +45°) after acetylation of the ethanolic extracts and determined its structure by spectral data and extensive chemical degradation. Subsequently, Andersen and Stonard (1979) succeeded in isolating the free phenol **49** as an unstable powder by rapid acid extraction followed by silica gel thin-layer chromatography. Ozonolysis of **50,** followed by reductive work-up, furnished 3,4,5-triacetoxybenzaldehyde (**51**) which was identical with an authentic sample. Hydrolysis of **50** with hydrochloric acid in acetonitrile gave the amide **52** which, by hydrogenolysis, afforded the debromo amide **53**. The latter was identical with an authentic sample of 2-acetamido-3-(indol-3-yl)propionamide. Comparison of its optical rotation with the literature value established the absolute configuration of **53**. The assigned structure **50** was confirmed by its reduction to **54,** which in turn was proved by synthesis.

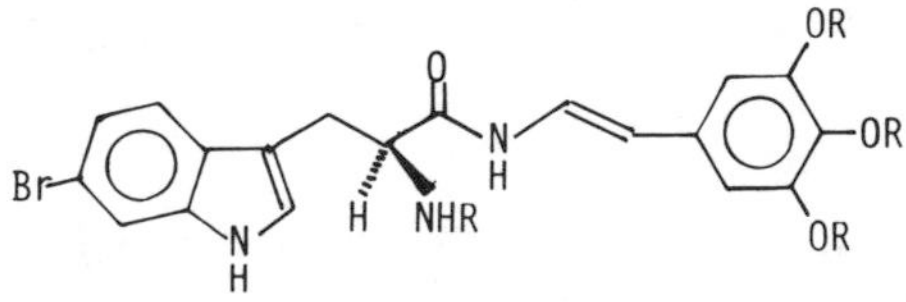

49   R = H
50   R = COMe

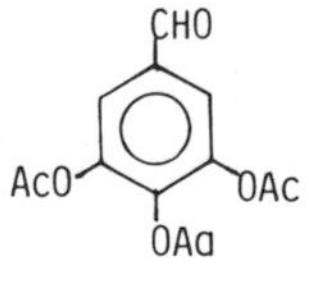

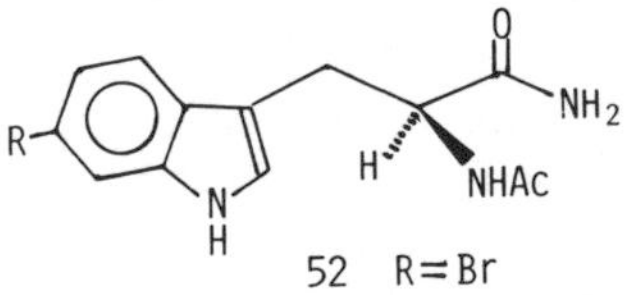

51

52   R = Br
53   R = H

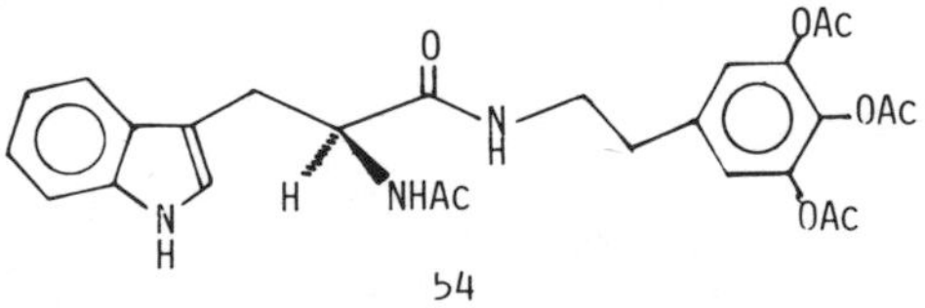

54

Many species in the sponge family Verongidae have been investigated for antibiotics and have provided over a dozen closely related compounds, which are believed to be metabolites of 3,5-dibromotyrosine (Minale *et al.*, 1976; Faulkner and Andersen, 1974). However, many of them have cyclohexadiene ring systems (e.g., **57**), and thus only a few of them retain a phenolic function. Stempien *et al.* (1973) isolated 3,5-dibromo-4-hydroxyphenylacetamide (**55;** mp 190°–191°) from methanolic extracts of the Caribbean sponge *Verongia archeri*. A similar metabolite, 2-hydroxy-3,5-dibromo-4-methoxyphenylacetamide (**56;** mp 174°–176°) has been found in the sponge *Psammoposilla purpurea* collected from Enewetak (Chang and Weinheimer, 1977). Both **55** and **56** were identified by comparison with authentic samples. Chang and Weinheimer (1977) secured an authentic specimen of **56** by the transformation of aeroplysinin-1 (**57**), another metabolite of the sponge, under the drastic conditions employed by Fattorusso *et al.* (1972). Phenol **56** showed anticancer activity (Chang and Weinheimer, 1977). Interestingly, antineoplastic activity has been observed in the related but much simpler *p*-hydroxyphenylacetamide (**58;** mp 174°–175°), a constituent of a sponge,

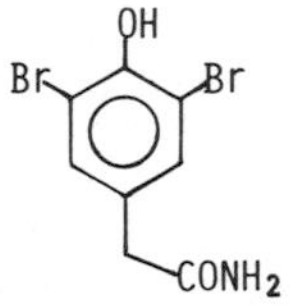

55 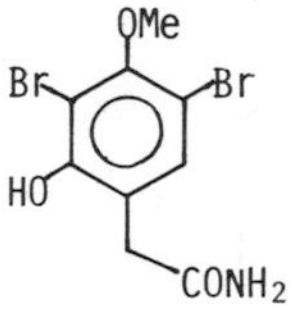   56 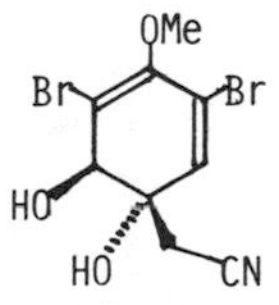   57 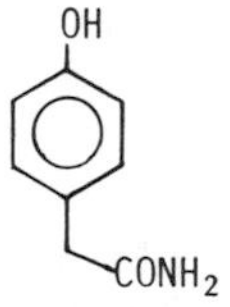   58

Anthosigmella varians (Schmitz *et al.*, 1977). Two rearranged dibromotyrosine metabolites have been found in a Californian sponge by Krejcarck *et al.* (1975). They isolated 2,4-dibromo-3,6-dihydroxy phenylacetamide [**59;** mp 170°–172° (dec)] and detected its monobromo analog (**60**) by GC–MS from the 2-propanol extract of *Verongia aurea*. The extract was concentrated and partially redissolved in ether, from which **59** crystallized on concentration. The structure was unambiguously established by X-ray crystallographic determination.

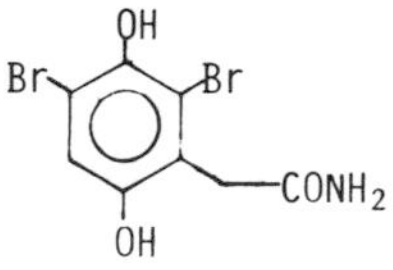

59 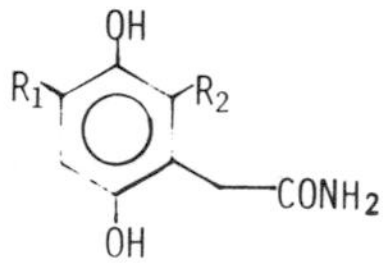   60 R_1=Br R_2=H or
            R_1=H R_2=Br

A related metabolite has been found in a red alga. Chantraine *et al.* (1973) treated the aqueous ethanol extract of *Halopytis incurvus* with diazomethane and isolated the methylated derivative **62** (mp 36°) of 3,5-dibromo-4-hydroxyphenylacetic acid (**61**) along with compound **78**, derived from 3,5-dibromo-4-hydroxyphenylpyruvic acid (**77**). The latter will be discussed in Section II,D. These bromophenols may be the immediate precursors in the biogenesis of their lower homologs, such as **21,** in some red algae.

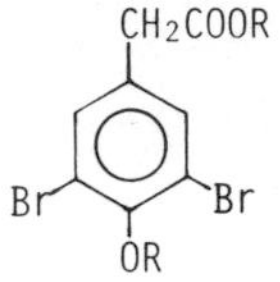

61   R = H
62   R = Me

D. C$_9$ Phenols

1. Phenolic Amino Acids

Halogenated tyrosines **(63–67)** have been found in a number of marine organisms, which include gorgonians (Drechsel, 1907; Mörner, 1913), sponges (Ackerman and Müller, 1941; Low, 1951), mollusks (Hunt and Breuer, 1971), and hemichordates (Barrington and Thorpe, 1963). Since iodotyrosines **(63, 64)** are the biological precursors of thyroid hormones, i.e., thyroxine and triiodothyronine, they seem to be relatively common in marine vertebrates and in some invertebrates which have thyroid function. However, from an evolutionary point of view, it is interesting to find thyroid hormone precursors and other halogenated tyrosines in lower animals, in which the existence of thyroid function has not been proved. 3-Bromo-5-chlorotyrosine **(67)** isolated from hydrolysates of scleroprotein of the mollusk *Buccinum undatum,* appears to be the only aromatic compound containing both chlorine and bromine atoms on a benzene ring.

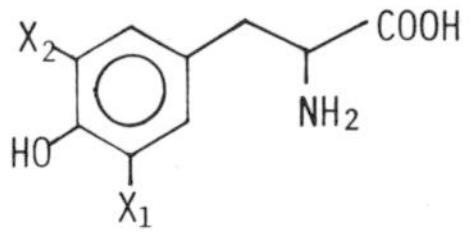

63   X$_1$ = X$_2$ = I
64   X$_1$ = I, X$_2$ = H
65   X$_1$ = X$_2$ = Br
66   X$_1$ = Br, X$_2$ = H
67   X$_1$ = Cl, X$_2$ = Br

Some tyrosine-derived metabolites of marine organisms include two epimers of majusculamide A **(68;** mp 96°–97°, [α]$_D$ +19.3°) and B **(69;** mp 102°–103°, [α]$_D$ +14.6°) and adenochromine A **(70),** B **(71),** and C **(72).**

Majusculamides have been isolated from the blue-green alga *Lyngbya majuscula*. Their structures, including absolute configuration, were established by X-ray crystallographic analysis of **69** (Marner *et al.*, 1977). Detailed discussion is given by Moore (1978) in Volume I of this series.

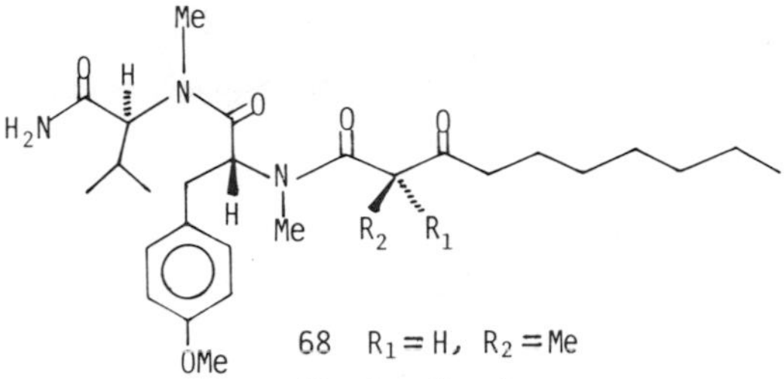

68   $R_1 = H$, $R_2 = Me$
69   $R_1 = Me$, $R_2 = H$

Adenochromines are the iron-chelating moieties of adenochrome, the iron(III)-containing pigment isolated from the branchial heart of *Octopus vulgaris* (Ito *et al.*, 1976; Prota *et al.*, 1977). Adenochrome is an inseparable mixture of closely related peptides, all of which are composed of an adenochromine plus two molecules of glycine. For details, including biosynthesis of adenochromines, Prota *et al.* (1977) should be consulted.

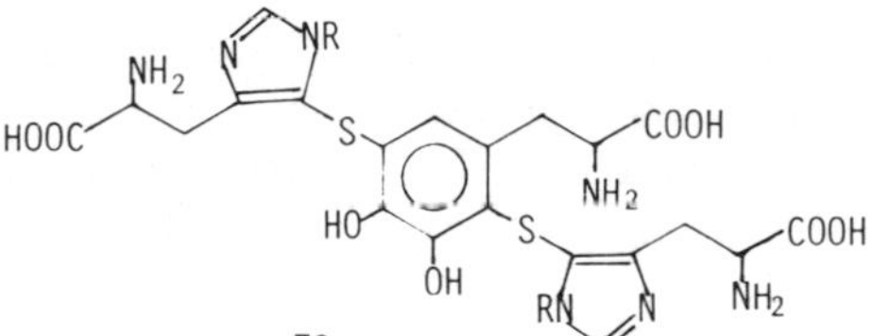

70

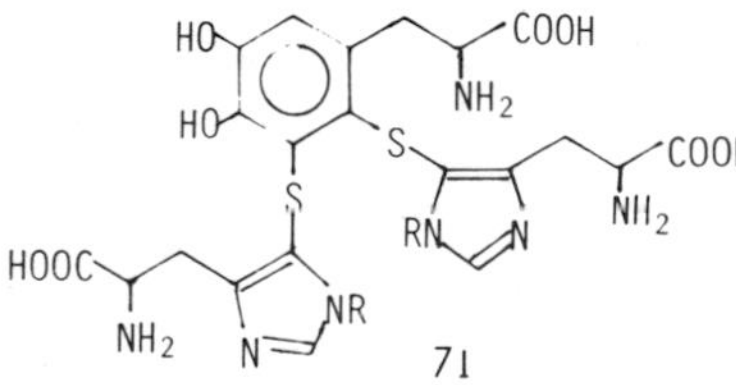

71

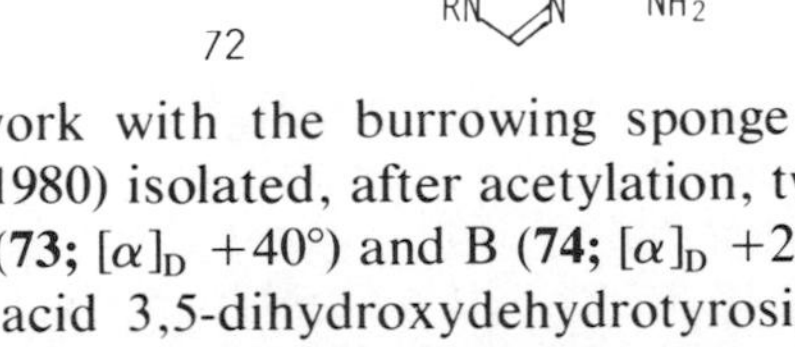

72

$R = H$, or Me

In their continuing work with the burrowing sponge *Cliona celata* Stonard and Andersen (1980) isolated, after acetylation, two novel tetrapeptides, celanamide A (**73**; $[\alpha]_D$ +40°) and B (**74**; $[\alpha]_D$ +22°), which contain the unique amino acid 3,5-dihydroxydehydrotyrosine. The structures of these peptides were established on spectroscopic and chemical grounds. It was proved, by employing (acetic anhydride)-d_6 as the acetylating agent of the extract, that the original sponge metabolites exist in their free phenolic forms (**75, 76**).

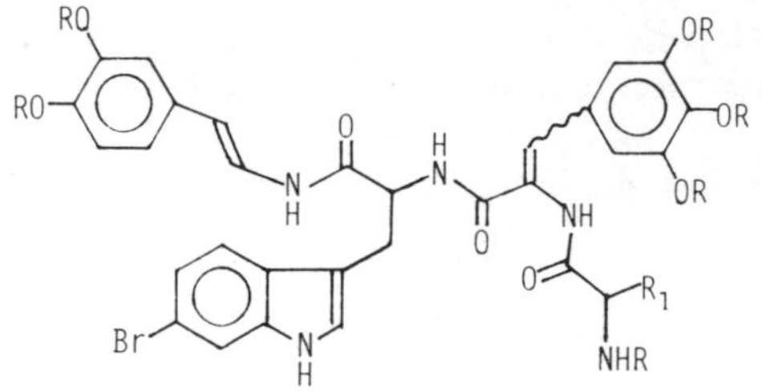

73   R = Ac, R_1 = $-CH_2CH(CH_3)_2$
74   R = Ac, R_1 = $-CH(CH_3)_2$
75   R = H, R_1 = $-CH_2CH(CH_3)_2$
76   R = H, R_1 = $-CH(CH_3)_2$

2. Other C_9 Phenols

As described in Section II,C, 3,5-dibromo-4-hydroxyphenylpyruvic acid (77; mp 248°) has been isolated as its trimethyl derivative (78; mp 96°) from the red alga *Halopytis incurvus* (syn. *H. pinastroides*) (Chantraine *et al.*, 1973). Structure 78 was deduced from spectral data, and the fact that it was derived from 77 was confirmed by preparing it from synthetic 77 and diazomethane. A related compound, 3-(4-hydroxyphenyl)-2-oximino-propanoic acid (79), has been isolated from the acetone extract of the sponge *Hymeniacidon sanguinea* (Cimino *et al.*, 1975b). Crude, unstable material obtained by separation on a column of ion-exchange resin was purified after methylation with diazomethane to yield trimethyl derivative 80, which was characterized by spectral analysis and confirmed by syn-

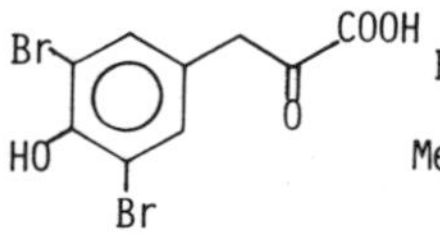

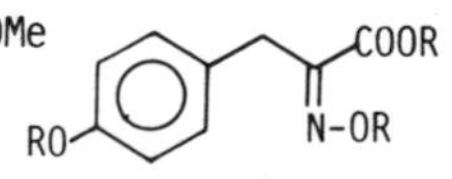

79   R = H
80   R = Me

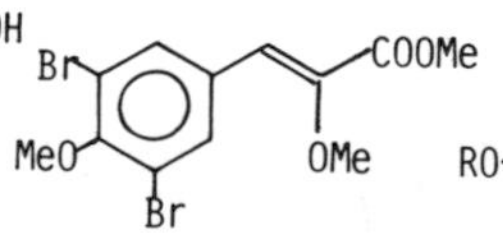

77

78

thesis. Moreover, the structure of the oxime 79 was definitely proved by catalytic reduction of the crude material to tyrosine. Cimino *et al.* (1975b) suggested that oximes such as 79 would be biogenetic precursors of a series of the metabolites, e.g., aeroplysinin-1 (57; Fattorusso *et al.*, 1972) and aerothionin (81; Moody *et al.*, 1972), obtained from the sponges of the family Verongidae (Scheme 2).

Recently Cariello *et al.* (1979) isolated caffeic acid (82; mp 200°) from the sea grass *Cymodocea nodosa* and its derivative, chicoric acid (83; mp 203°–205°, [α]_D −370°) from both *C. nodosa* and *Posidonia oceanica*. Saponification of 83 yielded caffeic acid (82) and (+)-tartaric acid. Chicoric acid from these marine plants is the optical antipode of the

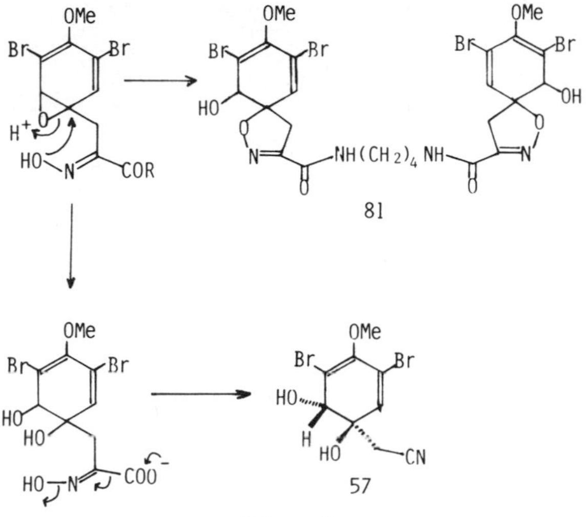

Scheme 2

compound isolated from the terrestrial plant *Chicorium intybus* (Scarpati and Oriente, 1958).

E. C₁₀ and Higher Phenols

One of the few antibiotics (for review, see Faulkner, 1978) obtained from marine microorganisms is 2-(2-hydroxy-3,5-dibromophenyl)- 3,4,5-tribromopyrrole (**84**), which was first isolated by Burkholder *et al.* (1966) from the marine bacterium *Pseudomonas bromoutilis,* and its structure was determined by Lovell (1966) by X-ray crystallographic analysis. It was immediately synthesized by Hanessian and Kaltenbronn (1966) as shown in Scheme 3. Later, Andersen *et al.* (1974) found the same antibiotics (**84**) along with tetrabromopyrrole (**85**), hexabromobipyrrole, *n*-propyl *p*-hydroxybenzoate (**30**) and *p*-hydroxybenzaldehyde (**19**) in a culture of *Chromobacterium* sp. Although biosynthetic origin of phenol **84** is by no means apparent, it has been placed in this section since co-occurrence of other metabolites in *Chromobacterium* sp. suggests that it could possibly

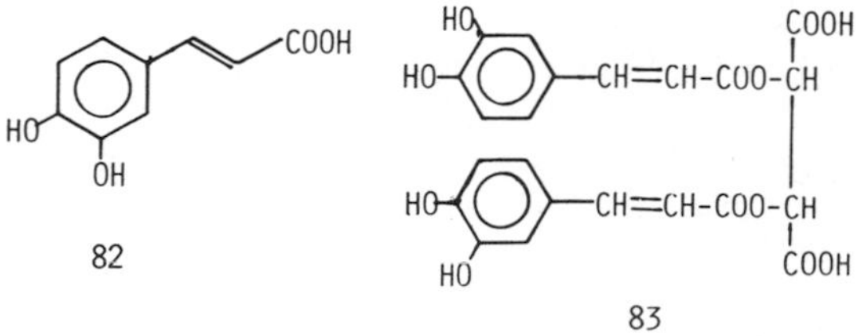

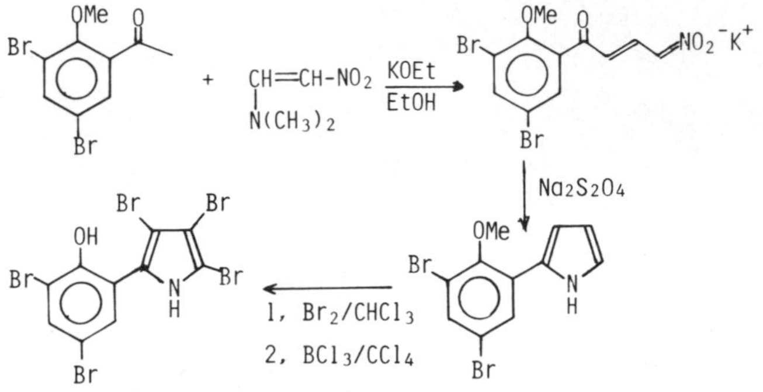

Scheme 3

be derived through condensation of **85** with one of the phenolic metabolites or their precursors.

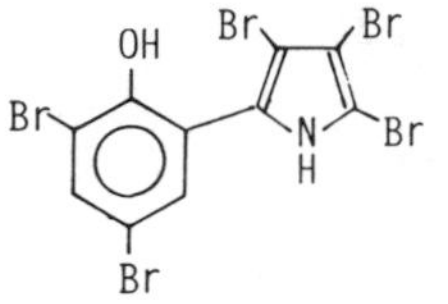

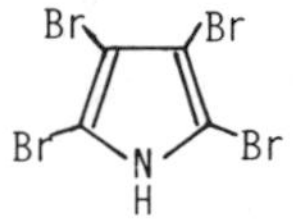

84   85

Bergmann and McAleer (1951) isolated fron the sponge *Spheciospongia vesparia* metanethole (**86**; mp 135°) which had been known (Baker and Enderby, 1940) as the dimer of anethole (**87**). The structure was confirmed by comparison with synthetic material. Since the compound was optically inactive, the authors suggested that it might be an artifact, presumably derived from anethole (**87**). Another suggested precursor was methylchavicol (**88**), which could rearrange to **87** under alkaline conditions that had been employed for the saponification of the original extract. Neither **87** nor **88** has been isolated from marine sponges, but these or other related substances have been suggested to be responsible for the aromatic odor often observed with some sponges (Bergmann and McAleer, 1951).

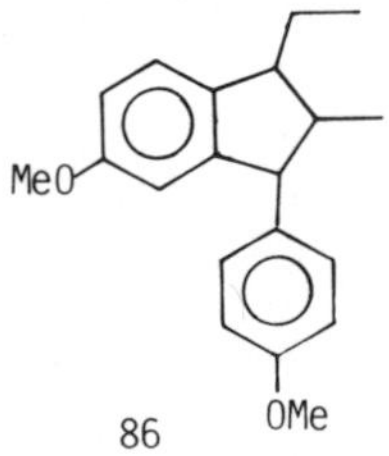

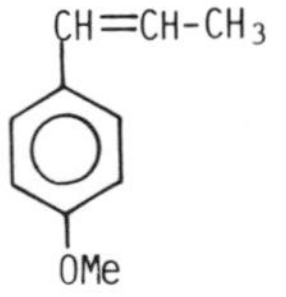

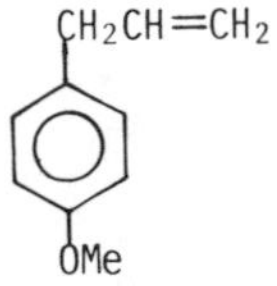

86   87   88

There are many phenolic compounds whose carbon skeletons could be represented by $C_6-C_3-(C_2)_n$ and therefore could be regarded as mixed metabolites derived by extending the shikimic acid-derived phenylpropanes (C_6-C_3) with acetate (C_2) units (see Geissman and Crout, 1969). The flavonoids $(C_6-C_3-C_6)$, which are widely distributed in terrestrial plants, are typical. Very few such phenols have so far been isolated from marine sources. Two of them, monoglucosides of quercetin (**89**) and isorhamnetin (**90**), were recently isolated as the peracetates from the marine phanerogame *Cymodocea nodosa* (Cariello *et al.*, 1979). These were identified by spectral and chemical methods. Upon hydrolysis, peracetylated **89** furnished glucose and quercetin, which was identified by comparison with an authentic sample. Similarly, hydrolysis of peracetylated **90** yielded glucose and an aglycone which, without identification, was treated with 30% potassium hydroxide solution to afford phloroglucinol (**5**) and 3-methoxy-4-hydroxybenzoic acid (**27**). From the results of alkaline hydrolysis of both **89** and **90,** the position of the glucose attachment was suggested to be either at C-5 or C-7. Markham and Porter (1969) had previously isolated a number of flavones from the green alga *Nitella hookeri*. These were C-glycosidic flavones related to the known vicenins (**91**) and lucenins (**92**), which had been isolated from a New Zealand tree, *Vitex lucens* (Seikel *et al.*, 1966).

89 R H
90 R Me

91 R H
92 R OH

The last substance to be included in this section is navenone C (**95**), one of the most interesting phenols from a biological point of view. Sleeper and Fenical (1977) have recently reported on the fascinating chemistry of marine opistobranch chemoreception. They succeeded in isolating from the pheromone secretion of the sea slug *Navanax inermis* the alarm pheromones, navenones A (**93**, B (**94**), and C (**95**), which induced an immediate alarm response to terminate the trail-following behavior of the animal. Of these pheromones, navenone C (**95**) (acetate, mp 135°–137°) is phenolic and appears to be a related metabolite to the class $C_6-C_3-(C_2)_n$, although nothing is known of the biogenesis of these compounds. However, since all three compounds have the same side chain, and since the pyridine moiety of **93** is perhaps derived from nicotinic acid, the phenolic moiety of **95** may likewise be derived from *p*-hydroxybenzoic acid or from benzoic acid. Such a biosynthesis, utilizing an Aryl—CO—SCoA rather than the more common Aryl—CH═CHCO—SCoA and extending the chain with acetate units, has been suggested for the formation of *Aniba rosaeodora* metabolites (see Geissman and Crout, 1969). Synthesis of all three navenones has been reported (Sakakibara and Matsui, 1979).

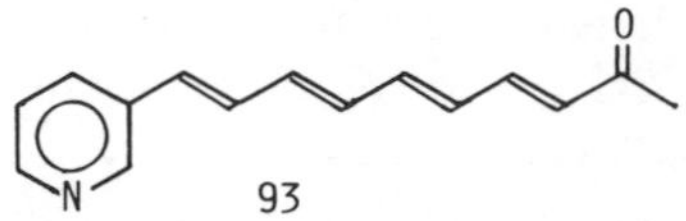

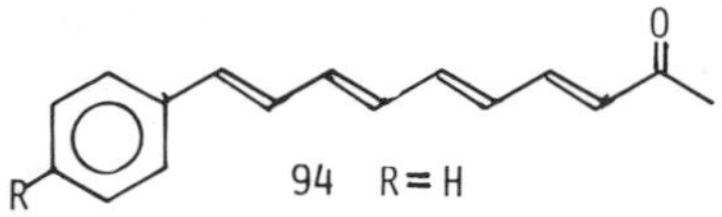

III. OLIGOPHENOLS

A. Diphenylmethane-Type Oligophenols

1. Dimers of Algal Phenols

As discussed in Section II, the members of the red algal family Rhodomelaceae are rich sources of simple bromophenols. Some bromophenols have been found to exist as dimers and trimers in these algae. Kurata and Amiya (1977) isolated from the red alga *Rhodomela larix* two dimeric phenols, 2,2′,3,3′-tetrabromo-4,4′,5,5′-tetrahydroxydiphenylmethane (**96**; mp 200°–201°) and 2,2′,3-tribromo-3′,4,4′,5-tetrahydroxy-6′-methoxymethyldiphenylmethane (**97**), which charred above 200° without melting and completely decomposed at ~540° as determined by differential thermal analysis. The structures of these phenols were assigned mainly by ¹H and ¹³C nmr determinations. The methyl ether **97** was suspected to be an artifact formed by methylation of **98** with methanol used for the extraction. Lundgren *et al.* (1979) synthesized **96** by condensation of 2,3-dibromo-4,5-dihydroxybenzyl alcohol (**33**) with catechol followed by bromination. Phenol **96** has also been identified from other members of the Rhodomelaceae, i.e., *Polysiphonia brodiaei* (Lundgren *et al.*, 1979), *P. nigrescens*, and *R. confervoides* (Pedersén, 1978). The Kurata group had previously reported a head-to-head dimer,

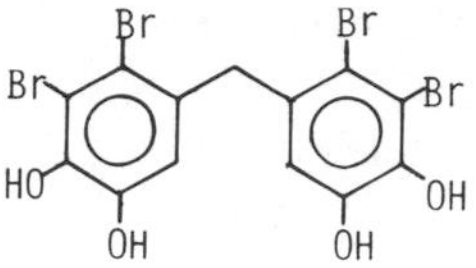

96 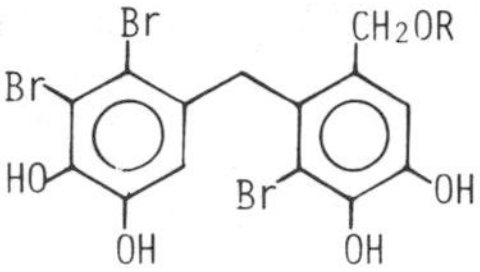   97   R   Me

98   R   H

3,3'-dibromo-4,4',5,5'-tetrahydroxybibenzyl (**99;** mp 205°–206°) from the methanolic extract of *P. urceolata* (Kurata *et al.*, 1976). The structure of **99** was elucidated by spectral and chemical means. Permanganate oxidation of the methylated derivative (**100**) afforded 3-bromo-4,5-dimethoxybenzoic acid (**101**).

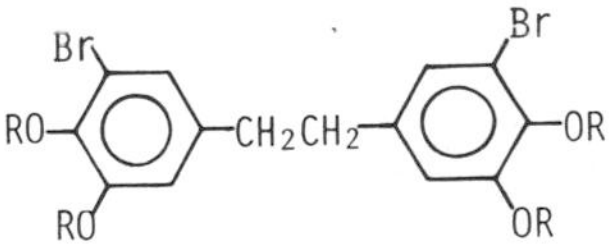

99   R = H   101

100   R = Me

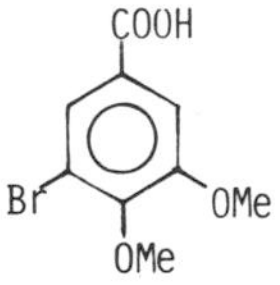

Chevolot-Magueur *et al.* (1976) isolated from the ethanolic extract of the red alga *Rytiphlea tinctoria* a diphenylmethane which consists of biosynthetically different aromatic rings. The compound which was isolated after treatment of the algal extract with diazomethane was identified as 2,3,3',5-tetrabromo-2',4,4',5,6'-pentamethoxydiphenylmethane (**102;** mp 129°) by spectral and X-ray single crystal determinations. The original phenol (**103**) is presumably derived in the alga by condensation of lanosol (**33**) of shikimic acid origin with acetate-derived 2,4-dibromophloroglucinol (**104**), both of which were isolated as permethylated derivatives from the alga.

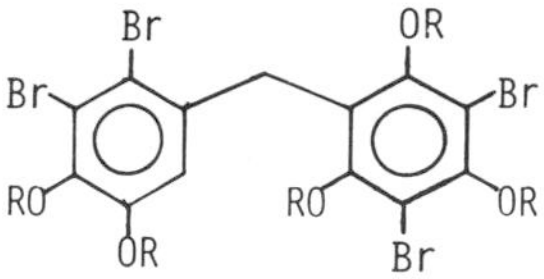

102   R = Me   104

103   R = H

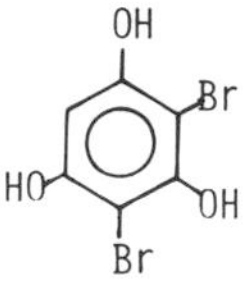

2. Cyclotribromoveratrylene

The red alga *Halopytis pinastroides* (syn *H. incurvus*) had previously been reported to contain the phenols **20, 35, 61,** and **77** (*vide supra*). More recently, Combaut *et al.* (1978a) obtained a cyclic trimer, cyclotribromoveratrylene (**105;** mp 222°) from the ethanolic extract of the same alga, which was treated with refluxing 1 *N* HCl and then with diazomethane before separation. Although the compound is rather novel, it appears to be an artifact formed by the condensation of 3-bromo-4,5-dihydroxybenzyl alcohol (**20**) during acid treatment. The structure of the compound was determined to be 10,15-dihydro-1,6,11-tribromo-2,3,7,8,12,13-hexamethoxy-5*H*-tribenzo[*a,d,g*] cyclononene (**105**) by spectral analysis and by comparison of its reduction (LiAlH$_4$) product with the known cycloveratrylene (**106**). The latter compound, a condensation

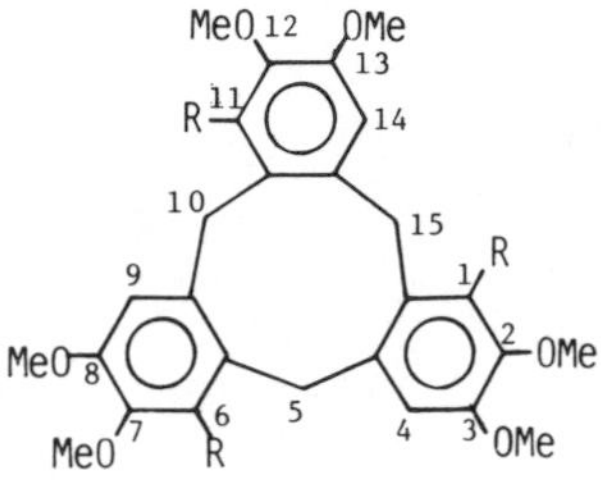

105   R = Br
106   R = H

product of veratrole with formaldehyde, has been shown to exist in the rigid crown conformation (**107**) by its ^{1}H nmr spectrum, which exhibits an AB system for the methylene protons (δ 4.64 and 3.48, *J* = 14 Hz) and remains unchanged up to 200° (Lindsey, 1965; Cookson *et al.*, 1968). However, unlike cyclotriveratrylene, an nmr spectrum of **105** showed the methylene signal at δ4.13 as a singlet (Combaut *et al.*, 1978b). This observation indicated that **105** exists in the flexible saddle conformation (**108**), which rapidly interconverts at room temperature among three equivalent structures. This conformation minimizes nonbonding interactions between the bulky bromine atoms and the hydrogen atoms at C-4,

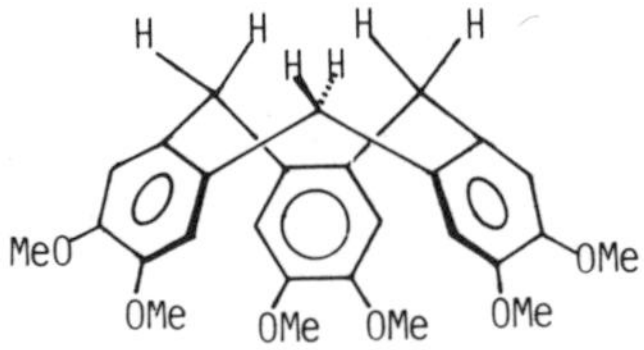

107
"Crown" conformation

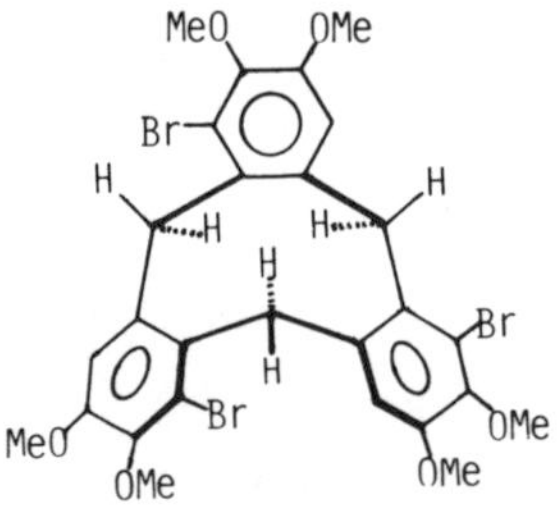

108
"Saddle" conformation

C-9, and C-14. The saddle conformation has also been observed with some other cyclotriveratrylenes in which C-1, C-6, and C-11 are occupied by bulky substituents (Manville and Troughton, 1973).

3. Thelepin and Related Compounds

Oligomeric phenols of the diphenylmethane type are not exclusively algal metabolites. Some similar substances have been found in a marine animal. We isolated from the annelid *Thelepus setosus* bis(3,5-dibromo-4-hydroxyphenyl)methane (**109**; mp 230°–232°), thelephenol (**110**; mp 183°–184°), and thelepin (**111**; mp 202°–203°) along with monomeric phenols **21** and **24** (Higa and Scheuer, 1974, 1975). The structure of **109** and **110** were deduced from spectral data and confirmed by synthesis. Bis(*p*-hydroxyphenyl)methane was prepared by condensation of phenol with formaldehyde and brominated to furnish **109**. Synthesis of **110** was performed by sodium borohydride reduction of the aldehyde **112** which

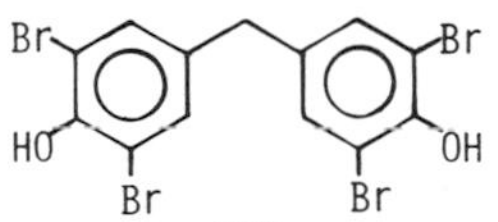

109

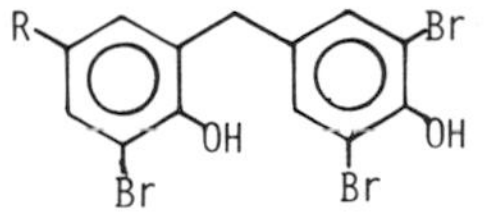

110   R = CH$_2$OH
112   R = CHO

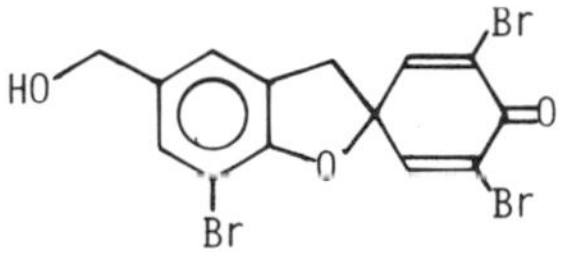

111

was prepared by the condensation of 3-bromo-4-hydroxybenzaldehyde with 3,5-dibromo-4-hydroxybenzyl alcohol in polyphosphoric acid. The structure of thelepin (**111**) was deduced by spectroscopic means and confirmed by chemical transformations. Catalytic reduction (Pd/C) of **111** furnished 2,4′-dihydroxy-5-methyldiphenylmethane (**113**) which was identified by comparing with a synthetic sample (Pummerer *et al.*, 1925). On the other hand, reduction of **111** with sodium borohydride afforded **110** and **114** in equal amounts. The dienol **114** was characterized by spectral analysis and confirmed by its ready conversion to **110** in methanolic hydrochloric acid (Scheme 4).

Thelepin has striking similarities to the well-known antifungal agent (for review, see Grove, 1963), griseofulvin (**115**), not only in its structure, but also in the level of antifungal activity. Thelephenol (**110**) appears to be the immediate precursor in the biogenesis of thelepin, in full analogy with griseophenone A (**116**), which has been shown to be the precursor (McMaster *et al.*, 1960) in the biogenesis of griseofulvin. Griseofulvin

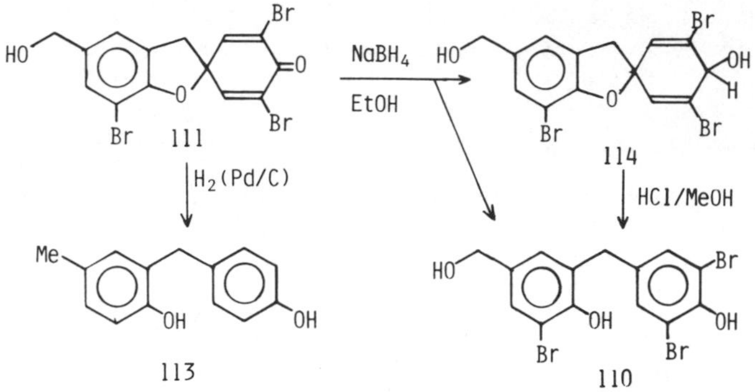

Scheme 4

(115), however, has been shown to be derived from acetate (Birch *et al.*, 1958), while thelepin (111) is in all probability a shikimic acid-related metabolite formed via simple phenols such as 3,5-dibromo-4-hydroxybenzyl alcohol (21).

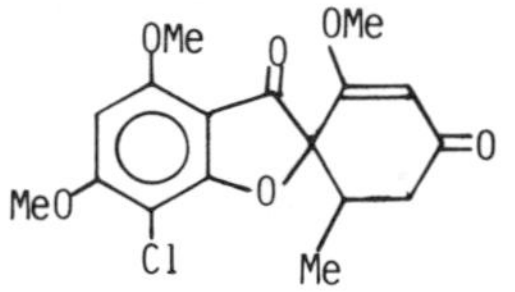

B. Diphenyl Ether-Type Oligophenols

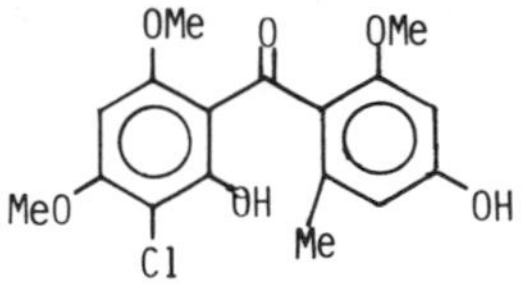

Sharma *et al.* (1970) reported five antibacterial (*p*-bromophenoxy)-bromophenols (118–122) from the sponge *Dysidea herbacea* collected in the Western Caroline Islands. Later Sharma and Vig (1972) reported the structures of two substances, including 122. These two were identified as 1-(4'-bromophenoxy)-6-bromo-2-hydroxybenzene (117; mp 95°–98°) and 1-(2',4'-dibromophenoxy)-2-hydroxy-4,5,6-tribromobenzene (122; mp 185°–186°) by spectral data and by hydrogenolysis that furnished known

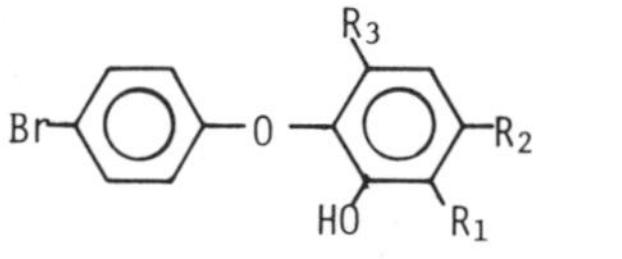

117 $R_1 = R_2 = H$, $R_3 = Br$
118 $R_1 = Br$, $R_2 = R_3 = H$
119 $R_1 = H$, $R_2 = R_3 = Br$

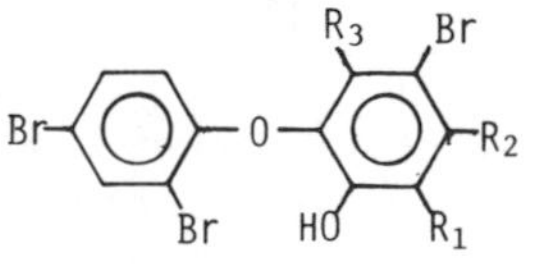

120 $R_1 = Br$, $R_2 = R_3 = H$
121 $R_1 = R_2 = Br$, $R_3 = H$
122 $R_1 = H$, $R_2 = R_3 = Br$

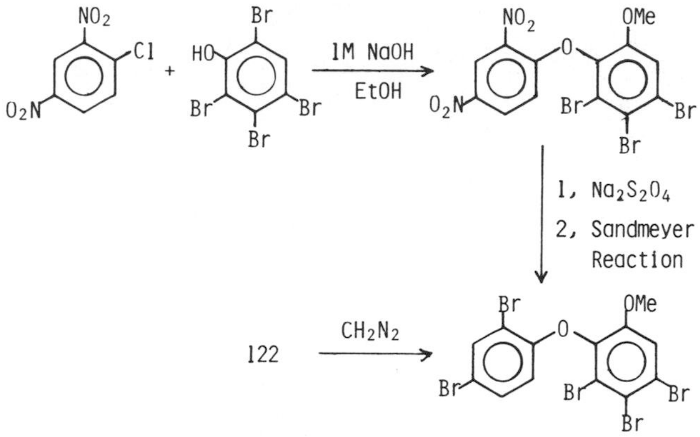

Scheme 5

o-phenoxyphenol (Lock, 1930). The structure of **122** was confirmed by synthesis of **123** as outlined in Scheme 5 (Sharma and Vig, 1972).

Since the work of Sharma *et al.* (1970), the sponge *D. herbacea* has attracted many workers and has yielded several nonphenolic, chlorinated metabolites (Hofheinz and Overhänsli, 1976; Kazlauskas *et al.*, 1977; Charles *et al.*, 1978). Kazlauskas *et al.* (1977) obtained, in addition to nonphenolic substances, a mixture of penta- and hexabromo derivatives of a diphenyl ether which they did not further separate.

From the Hawaiian acorn worm *Ptychodera flava laysanica* we isolated two dimers and a trimer of the bromohydroquinones encountered already in Section II,A (Higa and Scheuer, 1977). These were identified by spectral data and chemical transformations as 2,2',3',4,5,5'-hexabromo-3,4',6-trihydroxydiphenyl ether [**124;** mp 240°–242° (dec)], 2,2',3',4,5'-penta-

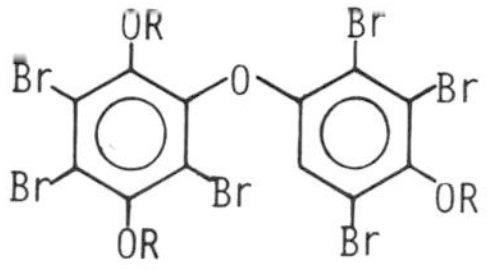

124   R = H
125   R = Me

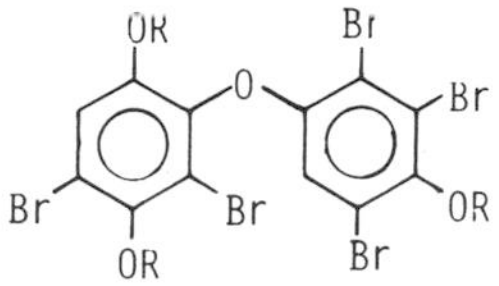

126   R = H
127   R = Me

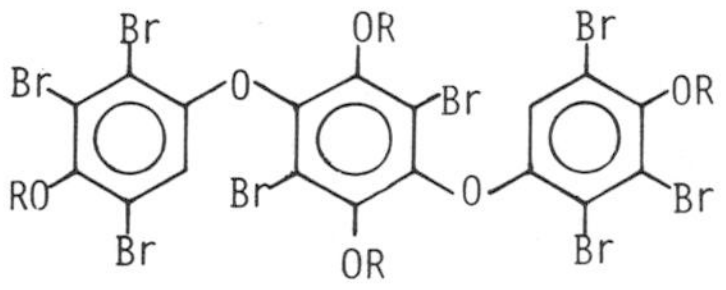

128   R = H
129   R = Me

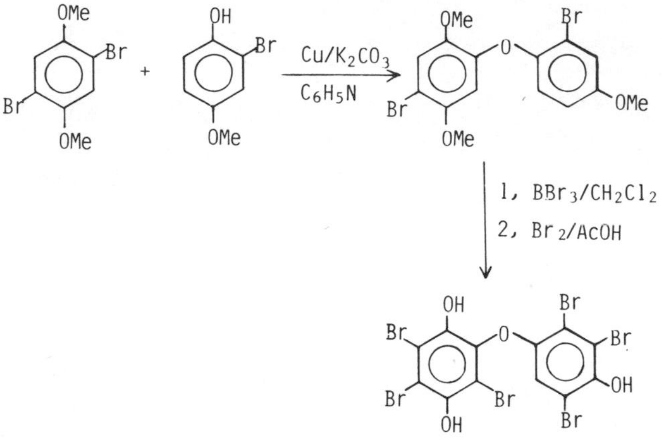

Scheme 6

bromo-3,4',6-trihydroxydiphenyl ether (**126**; 212°–215°), and 2,5-bis(4'-hydroxy-2',3',5'-tribromophenoxy)-3,6-dibromohydroquinone [**128**; mp 293°–295° (dec)]. Compound **124** was synthesized as outlined in Scheme 6. Hydrogenolysis (Pd/C) of **129,** derived from **128** by diazomethane treatment, yielded 2,5-bis(4-methoxyphenoxy)-1,4-dimethoxybenzene (**130**), which was confirmed by synthesis. The synthesis was accomplished by the condensation of 2,5-dibromo-1,4-dimethoxybenzene with two equivalents of 4-methoxyphenol in the presence of cuprous oxide in refluxing collidine.

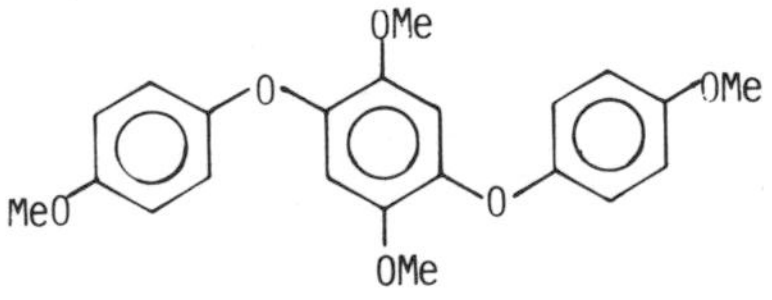

130

Conformational consideration provided an important clue for the position of the aromatic hydrogen in **124.** As shown by a number of related systems (Montaudo *et al.*, 1971), in the most stable conformation of diphenyl ethers bearing three bulky substituents ortho to the ether linkage the ring bearing ortho hydrogen is perpendicular to and bisects the plane of the other ring and the ortho hydrogen is near this plane. Therefore, the ortho hydrogen is nicely situated in the shielding region of the other aromatic ring as shown in **131.** Upfield shifts observed with the aromatic protons of **125** (δ6.59), **127** (δ6.52), and **129** (δ6.63) in comparison with those of tribromohydroquinone dimethyl ether (δ6.97) and other related

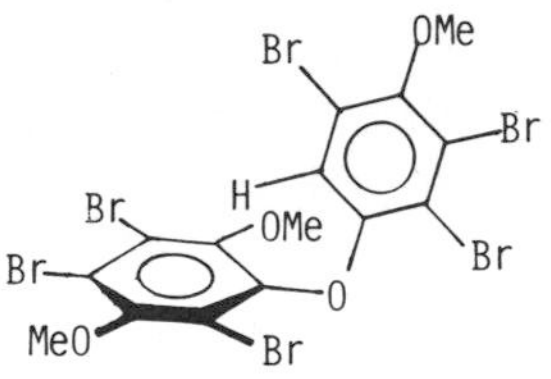

131

compounds constitute convincing evidence that these hydrogens are situated at positions ortho to the ether linkages.

C. Phlorotannins

Since Crato (1892) first suggested that the physodes, subcellular bodies of brown algae, contain phloroglucinol (5) and its derivatives, the occurrence of phloroglucinol has been demonstrated in the hydrolysates of *Ecklonia cava* (Takahashi, 1931), *Cystophyllum hakodatense* (Shirahama, 1942), *Sargassum ringgoldianum* (Ogino and Taki, 1957), and *Fucus vesiculosus* (Craigie and McLachlan, 1964). More recently, Glombitza *et al.* (1973) demonstrated the presence of phloroglucinol in 17 species of brown algae. Since then, the extensive work of the German group with brown algae has led to the isolation of no fewer than 20 oligomers of phloroglucinol, many of which have the properties of tannin. For this reason, Glombitza (1977) suggested a general term "phlorotannins" for this class of phenols. Glombitza (1977) further classified phlorotannins into four structural types: fucols, phlorethols, fucophlorethols, and fuhalols. Since these phenols are highly hydroxylated compounds and usually extremely unstable, they are most conveniently isolated and handled after acetylation or methylation. The German group has isolated most of them as peracetyl derivatives. For detailed discussion, reviews by Glombitza (1977, 1979) should be consulted.

1. Fucols

Fucols are biphenyl-type oligomers in which each molecule of phloroglucinol is connected by ring to ring C—C bonds. Difucol or 2,2',4,4',6,6'-hexahydroxybiphenyl (132) has been isolated from *Fucus vesiculosus* (Glombitza *et al.*, 1975a; Ragan and Craigie, 1976) and several other brown algae (Glombitza *et al.*, 1976b, 1977a, 1978b). The structure of 132 was confirmed by synthesis (Ragan and Craigie, 1976). Trifucol nonaacetate (133) has been isolated from *F. vesiculosus* and *Bifurcaria bifurcata* (Glombitza *et al.*, 1975a, 1976b). Tetrafucol dodecaacetate, a tetraphenyl derivative, has been isolated from *F. vesiculosus* (Glombitza *et al.*,

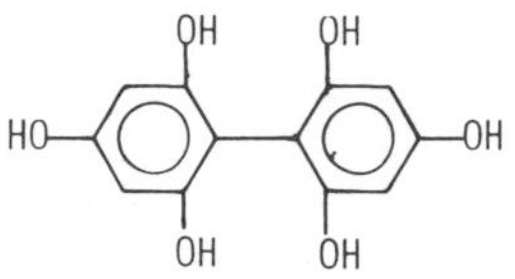

132 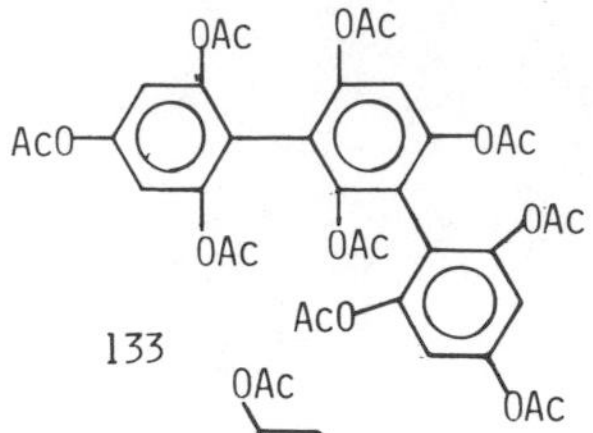   133

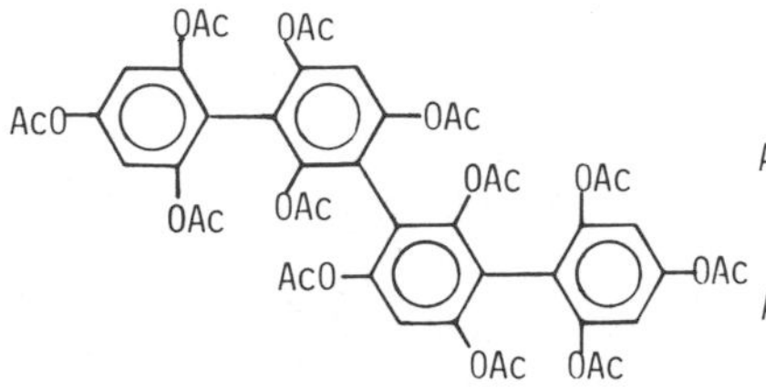

134 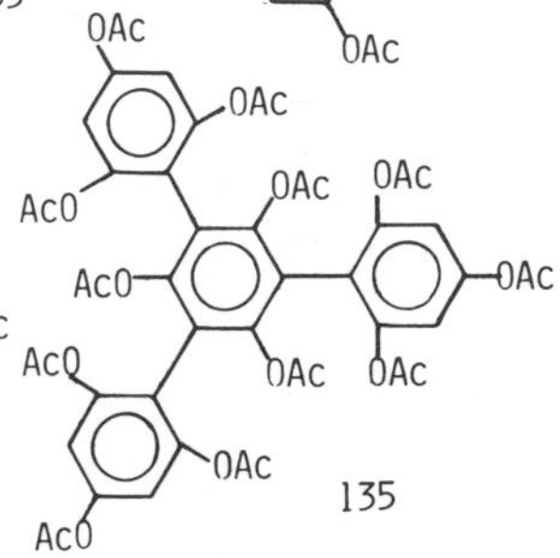   135

1975a). The authors suggested that it was a mixture of two isomeric structures **134** and **135,** none of which has yet been confirmed.

2. Phlorethols

Phlorethols are the oligomers in which phloroglucinol moieties are linked by ether linkages. Diphlorethol pentaacetate was isolated from the acetylated extract of *Cystoseira tamariscifolia* (Glombitza *et al.*, 1975b). The assigned structure was 2,3',4,5',6-pentaacetoxydiphenyl ether (**136;** mp 113°–114°) on the basis of spectral data. It was confirmed by synthesis from 2,4,6-trimethoxybromobenzene and 3,5-dimethoxyphenol. Compound **136** has been isolated from several other species of brown algae (Glombitza *et al.*, 1976a,b, 1977a, 1978a). The alga *Laminaria ochroleuca* which yielded **136** also contains its homolog, triphlorethol-C heptaacetate, or 1-acetoxy-3,5-bis(2,4,6-triacetoxyphenoxy)benzene (**137;** mp 68°–71°) (Glombitza *et al.*, 1976a) and a chlorinated derivative of **137,** chlorotriphlorethol-C heptaacetate (**138**) (Glombitza *et al.*, 1977b). Chlorination of **137** did not yield **138,** but rather, isomeric monochloro and dichloro derivatives. Chlorinated derivatives of diphorethol and higher homologs have also been detected by mass spectrometry.

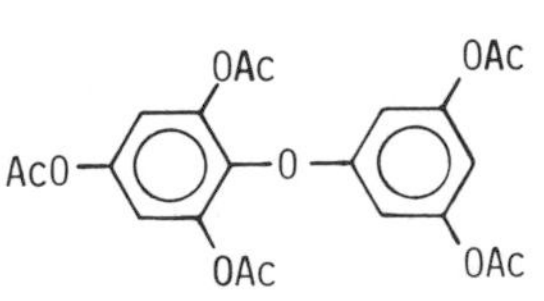

136

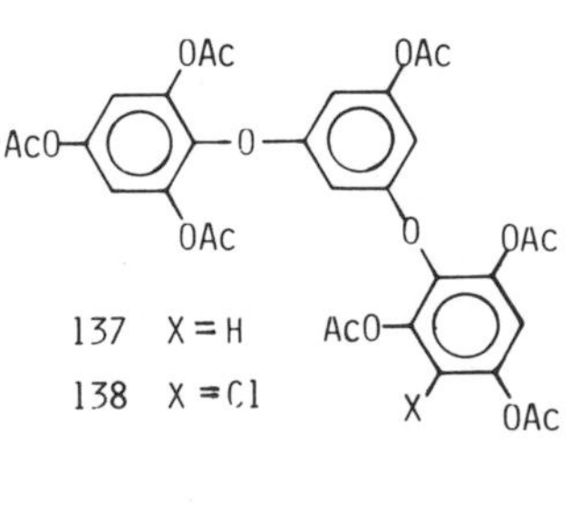

137 X = H
138 X = Cl

3. Fucophlorethols

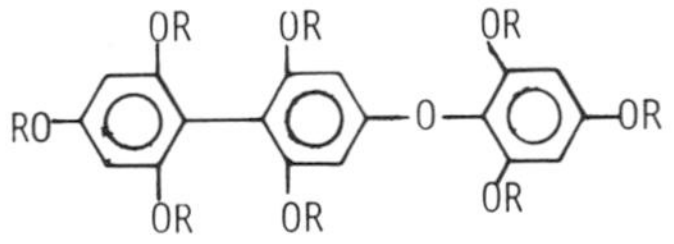

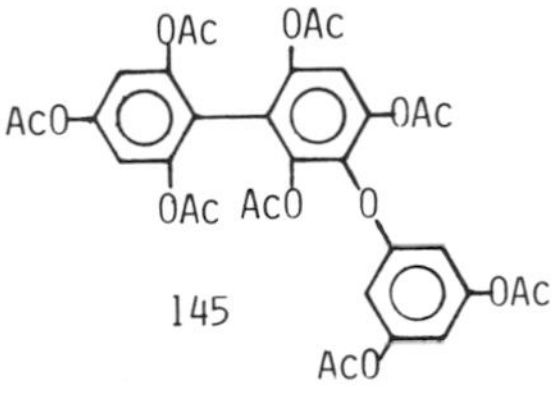

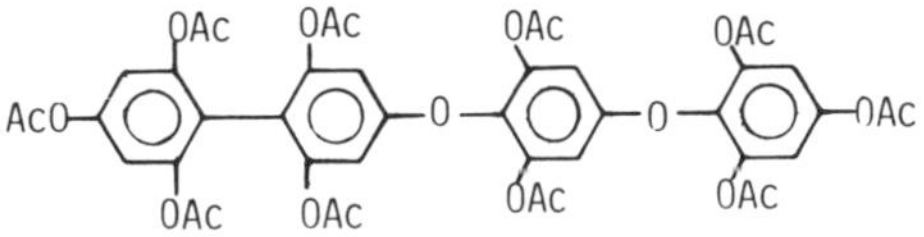

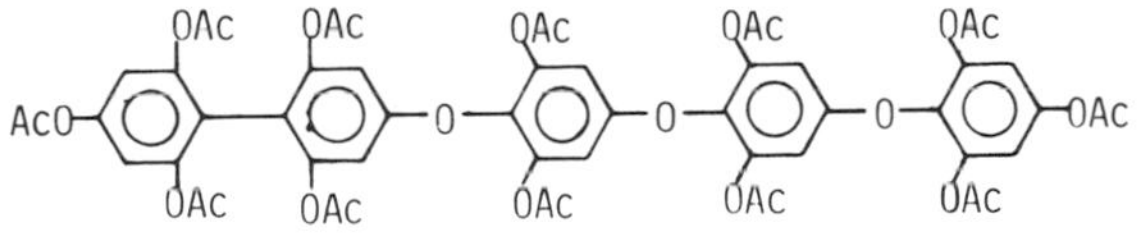

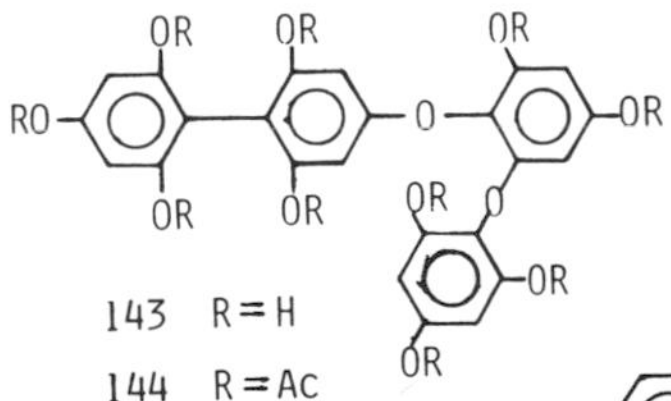

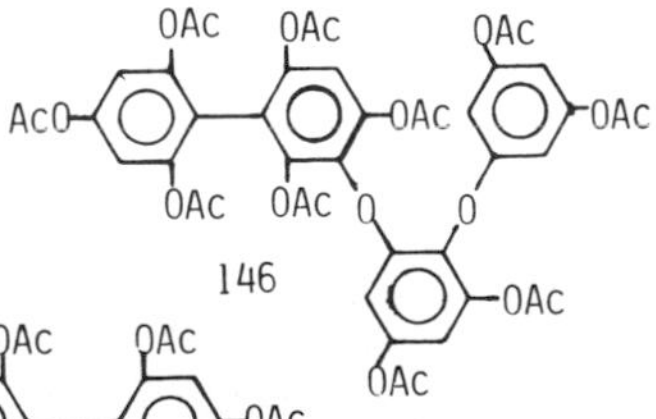

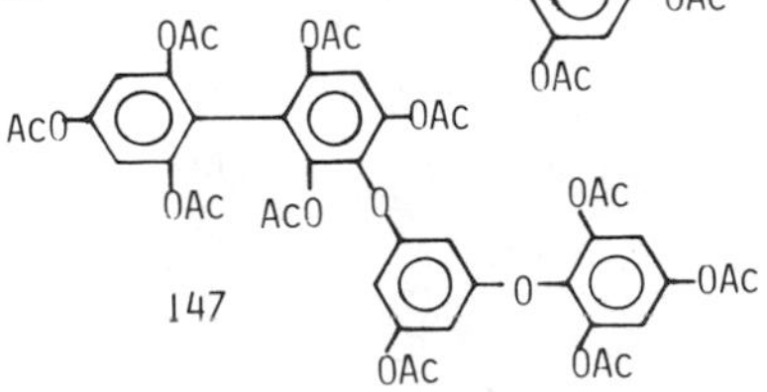

The third type of phlorotannins is the mixed forms of fucols and phlorethols. All the compounds found to date are those composed of one unit of difucol (**132**) with one to three molecules of phloroglucinol connected by ether linkages. Glombitza *et al.* (1977c) isolated from the acetylated extract of *Fucus vesiculosus* fucophlorethol-A octaacetate (**140**; mp 75°–83°) fucodiphlorethol-A decaacetate (**141**; mp 86°–94°) and fucotriphlorethol dodecaacetate (**142**; mp 105°–111°) while Craigie *et al.* (1977) isolated from the same alga two free phenols, fucophlorethol-A (**139**) and the angular isomer (**143**) of fucodiphlorethol, which confirmed the occurrence of such free phenols in the alga. These were converted to peracetyl derivatives (**140, 144**) and characterized by ^{1}H and ^{13}C nmr spectroscopy. Other related compounds include fucophlorethol-B octa-

139 R = H

140 R = Ac

145

141

142

143 R = H

144 R = Ac

146

147

acetate (**145**; mp 151°) and fucodiphlorethol-B decaacetate (**146**; mp 105°) from *Cystoseira baccata* (Glombitza *et al.*, 1978b), and a fucodiphlorethol (**147**) from *Laminaria ochroleuca* (Glombitza, 1977). No supporting data are available for the structure **147**.

4. Fuhalols

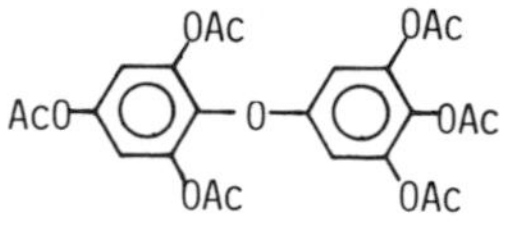

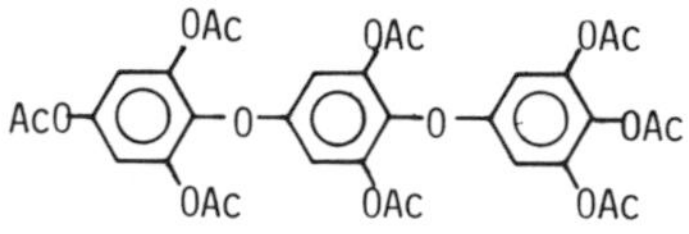

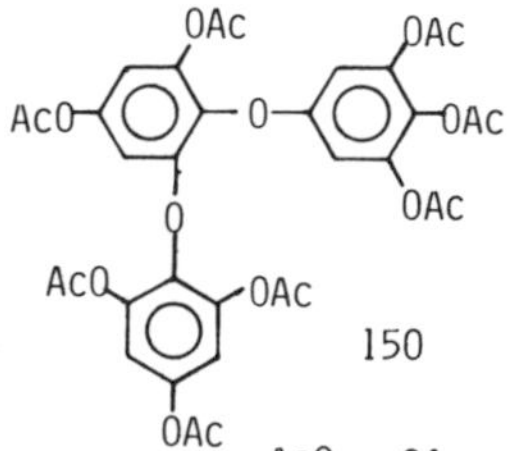

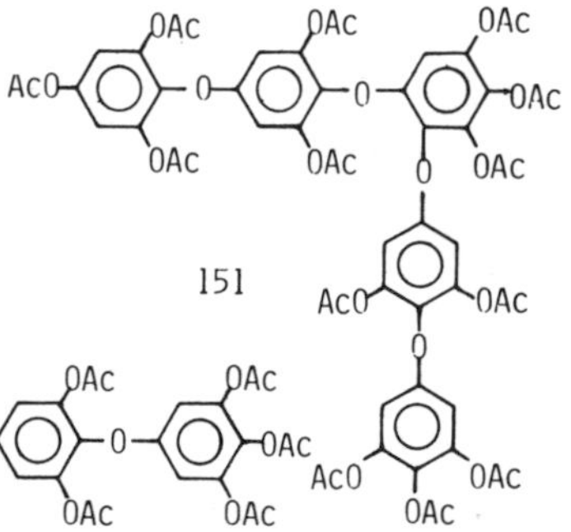

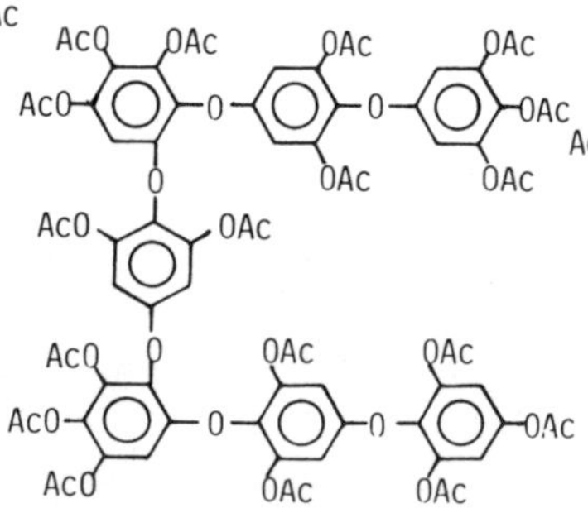

The fourth structural type is similar to the phlorethols, but bears one or more additional hydroxyl groups. The simplest fuhalol, bifuhalol hexaacetate (**148**; mp 184°–186°), was first isolated from *Bifurcaria bifurcata* and characterized by spectroscopic means (Glombitza and Rösner, 1974). Since then it has been isolated from three other species (Glombitza *et al.*, 1975b, 1978a,b). Trifuhalol-A octaacetate (**149**; mp 117°–123°) has been isolated from *Halidrys siliquosa* (Glombitza and Sattler, 1973), *B. bifurcata*, and *Sargassum muticum* (Glombitza *et al.*, 1976b, 1978a). The structures of both **148** and **149** were confirmed by synthesis with a modified Ullman reaction (Sattler and Glombitza, 1975). Other identified homologs include trifuhalol-B octaacetate (**150**) (Glombitza *et al.*, 1978a), pentafuhalol tridecaacetate (**151**; mp 105°–130°), (Glombitza *et al.*, 1976b), and heptafuhalol octadecaacetate (**152**; mp 112°–132°) (Glombitza *et al.*, 1976b; Sattler *et al.*, 1977).

IV. ACETATE-DERIVED PHENOLS

A large number of phenolic substances are biosynthetically derived via an acetate–malonate pathway, e.g., the phlorotannins described earlier, and phenolic naphthoquinones and anthraquinones, which are not included in this discussion. Phloroglucinol and 2,4-dibromophloroglucinol (**104**) have already been described in the preceding sections as acetate-derived simple phenols from marine sources. An additional example is 1,2,3,5-tetrahydroxybenzene 2,5-disulfate ester (**153**) which was isolated

153

from the brown alga *Ascophyllum nodosum* (Jensen and Ragan, 1978). A sample of **153,** although not rigorously pure, was obtained from the aqueous acetone extract of the alga through various treatments including preparative paper electrophoresis and Sephadex G-15 column chromatography and was characterized by spectral analysis. The structure was confirmed by synthesis of **153** as the dipotassium salt. The phenol **153,** which is presumably derived from phloroglucinol, was suggested to be the phenolic precursor for the formation of the marine "Gelbstoff" in the exudates of *A. nodosum*.

Gregson *et al.* (1977) have isolated a series of polyketide-derived resorcinols (**154–157**) and a phloroglucinol (**158**) from the brown alga *Cystophora torulosa*. These substances have been reviewed by Moore (1978) in Volume I of this series.

Another array of phenols, which may be regarded as products of an acetate–malonate pathway, are aplysiatoxin (**159**) and related compounds (**160–166**). The first two were isolated by Kato and Scheuer (1974, 1975) from the sea hare *Stylocheilus longicauda*. The structures of these fascinating compounds were determined by extensive degradation as well as the use of spectral techniques (Kato and Scheuer, 1974, 1975, 1976). For discussion of the structural elucidation and chemistry the reader is referred to the review by Moore (1978) in Volume I of this series.

In the course of anticancer studies on several species of blue-green

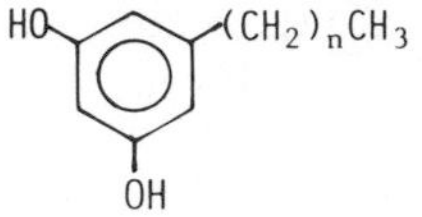

154   n = 12
155   n = 14

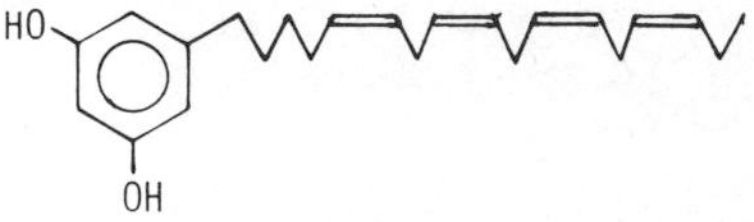

156

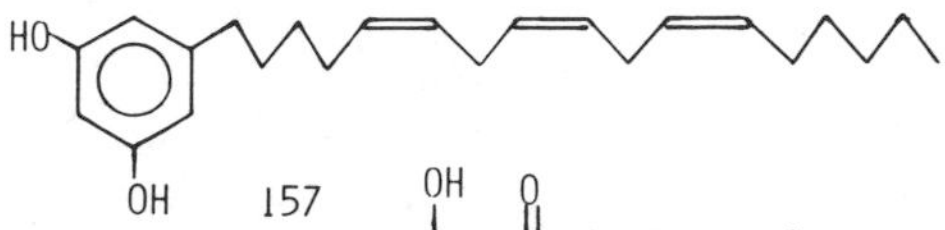

157

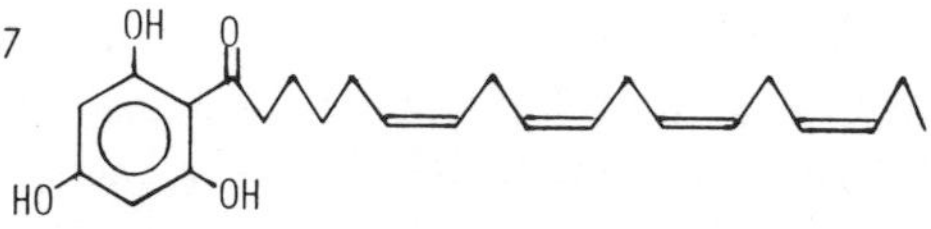

158

algae collected at Enewetak Atoll, Mynderse *et al.* (1977) isolated de-bromoaplysiatoxin (**160;** mp 105.5°–107°, $[\alpha]_D$ +60.6°) as an active component of *Lyngbya gracilis*. Subsequently, Mynderse and Moore (1978) isolated from a mixture of the blue-green algae *Oscillatoria nigroviridis*

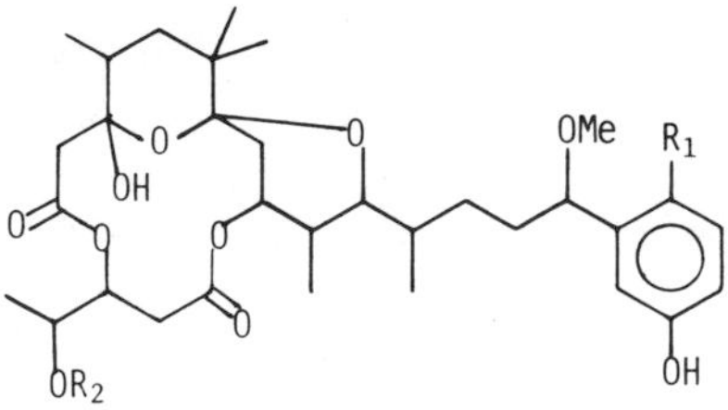

159   $R_1 = Br$, $R_2 = H$
160   $R_1 = R_2 = H$
161   $R_1 = Br$, $R_2 = Ac$
162   $R_1 = H$, $R_2 = Ac$

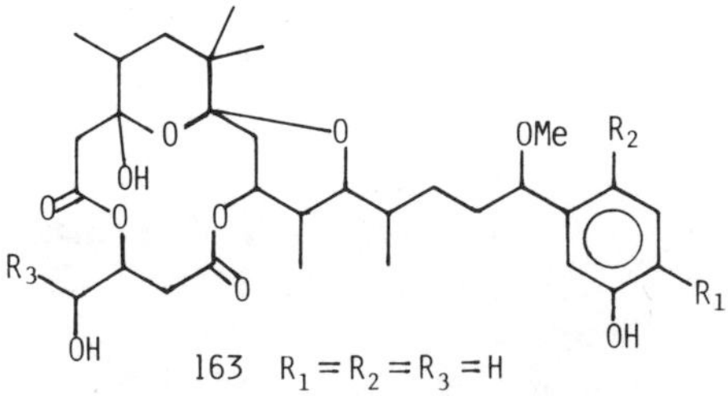

163   $R_1 = R_2 = R_3 = H$
164   $R_1 = R_3 = H$, $R_2 = Br$
165   $R_1 = R_2 = Br$, $R_3 = H$
166   $R_1 = R_2 = Br$, $R_3 = Me$

and *Schizothrix calcicola* oscillatoxin A (**163**; 31-nordebromoaplysia-toxin), $[\alpha]_D$ +67 ± 10°, and related minor components (**164–166**), all of which were identified by comparison of ^{1}H (360 MHz) and ^{13}C nmr spectra with those of debromoaplysiatoxin (**160**). Debromoaplysiatoxin was related to the cause of "swimmers' itch," a contact dermatitis produced by *Lyngbya majuscula* (Mynderse *et al.*, 1977). It also clarified the dietary origin of the aplysiatoxins in the sea hare *S. longicauda* (Scheuer, 1977).

A phenolic ether containing a bromoallene moiety has been isolated as one of the fish antifeedant constituents of the sea hare *Aplysia brasiliana* found in Florida (Kinnel *et al.*, 1977). The structure of the compound panacene (**167**; oil, $[\alpha]_D$ +382°) was assigned mainly by ^{1}H and ^{13}C nmr data. Absolute configuration of the allenic moiety was tentatively assigned by the application of Lowe's rule (Lowe, 1965). Panacene (**167**) is closely related to the class of halogenated unbranched C_{15} compounds having a terminal enyne group, most of which have been obtained from the algal genus *Laurencia* and often from sea hares which graze upon the algae (for review, see Moore, 1978).

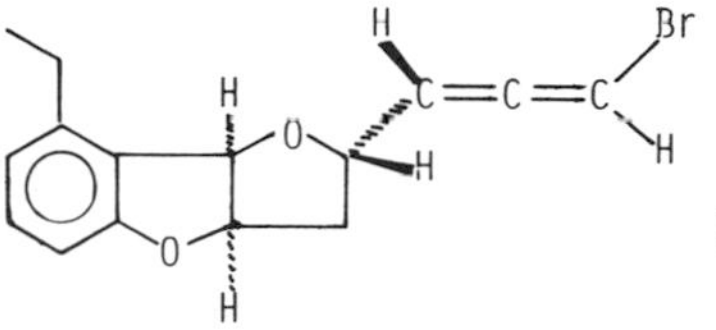

167 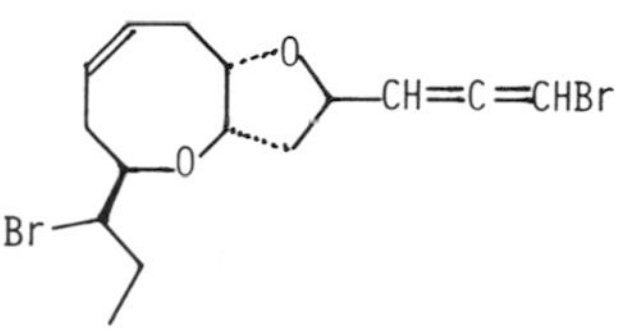   168

Biosynthesis of panacene has been proposed by a route involving the attack of bromine cation on the terminal acetylenic carbon followed by ring closure as shown in Scheme 7. A question has remained as to whether this process occurs in an alga or in the sea hare. The recent discovery of another closely related bromoallene, although not phenolic, favors the formation of allenic moiety in the alga. The second bromoallene, laurallene (**168**), was isolated from the red alga *Laurencia nipponica* collected in Hokkaido, Japan (Fukuzawa and Kurosawa, 1979).

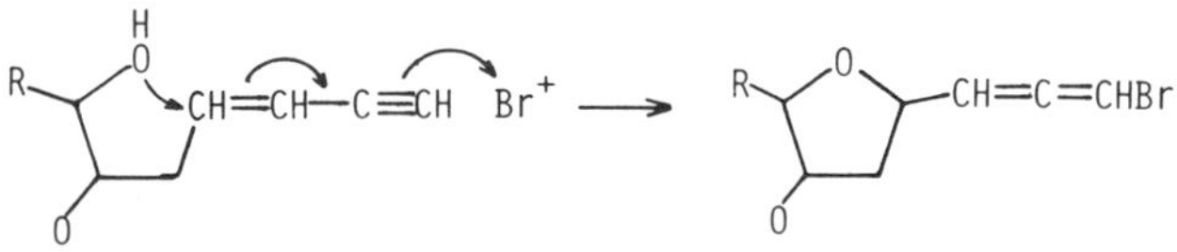

Scheme 7

V. MEVALONATE-DERIVED PHENOLS

Some 20 phenols and phenolic ethers, which are embodied in the structural framework of sesquiterpenes, are known from marine sources.

They may be classified into three types: α-curcumene, calamenene, and cuparene. The former two types have been obtained from gorgonians, while the latter are obtained from red algae and sea hares.

A. α-Curcumenes and Calamenenes

McEnroe and Fenical (1978) isolated curcuphenol (169; oil, $[\alpha]_D$ $-7.0°$), curcuhydroquinone (170; oil $[\alpha]_D$ $-21°$), and the corresponding ben-zoquinone, curcuqinone as the compounds responsible for the antibacte-rial activity of the extracts of the gorgonian *Pseudopterogorgia rigida*. In addition, several nonphenolic inactive sesquiterpenes including the parent hydrocarbon, α-curcumene (171), were isolated from the animal. Signif-

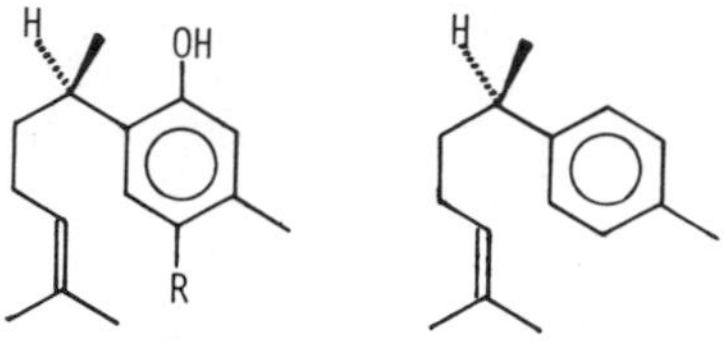

169   R = H   171
170   R = OH

icantly, the yield of hydroquinone 170 amounted to 18% of the extract. Structures of the phenolic components were based on spectral data and by reduction of the mesylate derivatives to yield the parent hydrocarbon 171. Comparison of the optical rotation with that of $(-)$-α-curcumene estab-lished their absolute configurations as R. Phenol 169 was synthesized in high overall yield (86%) as outlined in Scheme 8 (McEnroe and Fenical, 1978).

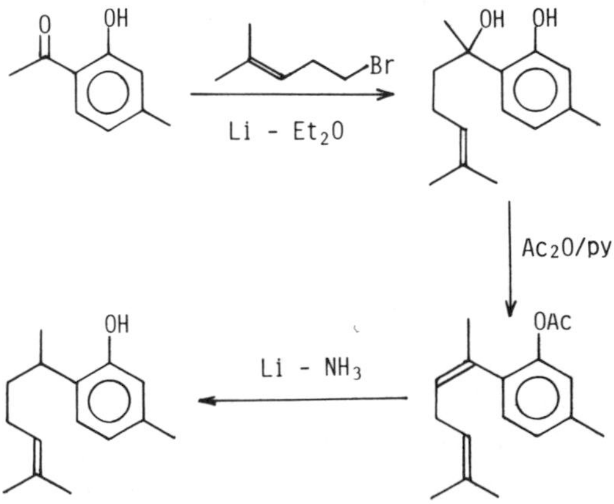

169

Scheme 8

From another gorgonian, *Subergorgia hicksoni,* collected in the Red Sea, Kashman (1979) isolated two oxygenated calamenenes. Structures of these compounds were determined by spectral data as 8-methoxycalamenene (**172**; oil, $[\alpha]_D$ +30°) and 5-hydroxy-8-methoxycalamenene (**173**; mp 84°, $[\alpha]_D$ +58°). Although the presence of a hydroxyl group at C-5 of **173** was proved by the formation of the monoacetate, **173** did not produce a color with ferric chloride. This failure was explained by the steric hindrance of **173**, in analogy with sterically crowded 2,6- disubstituted phenols, which do not undergo complexation with ferric chloride (see Tsuruta and Mukai, 1968). The stereochemistry of both terpenes remains unresolved.

172 R=H
173 R=OH

B. Cuparene-Related Phenols

Since Yamamura and Hirata (1963) first reported aplysin (**180**), debromoaplysin (**181**), and aplysinol (**182**), which were isolated from the sea hare *Aplysia kurodai,* a number of related sesquiterpenes have been found in the red algal genus *Laurencia* and in some sea hares which graze on the algae. Many of these compounds can be interconverted by simple treatments. The structural elucidations, chemical transformations, and syntheses have been thoroughly reviewed by Scheuer (1973) and by Martin and Darias (1978) in Volume I of this series. Sims *et al.* (1978) in Volume II have discussed ^{13}C nmr spectra of many of these compounds (**174, 175, 177, 178, 180, 181, 188, 189**). Thus, in this review I have listed all structures isolated to date, but discussion has been limited to those which have not been covered in previous reviews. Those covered in previous reviews include compounds **174–177, 180–182,** and **185**. Compound **185**, 2-(bromomethyl)-2,3,4,5-tetrahydro-5,8,10-trimethyl-2,5-methano-1-benzoxepin, had formerly been known as a product of the spontaneous transformation of laurenisol (**176**), which was obtained from *L. nipponica* (Irie *et al.*, 1969a). It (**185**) has now been isolated from *L. glandulifera* along with other cyclic ethers, 2-(dibromomethyl)-2,3,4,5-tetrahydro-5,8,10-trimethyl-2,5-methano-1-benzoxepin (**186**; mp 125°–126°, $[\alpha]_D$ +79.0°) and 7-bromo-2-(bromomethyl)-2,3,4,5-tetrahydro-5,8,10-trimethyl-2,5-methano-1-benzoxepin (**187**; mp 86°–87°, $[\alpha]_D$ +22°) (Suzuki and Kurosawa, 1976). The structures of **186** and **187** were assigned by comparing spectral data with those of **185**; the structure of **187** was further confirmed by treating **185** with bromine in acetic acid.

Irie *et al.* (1969b) had previously isolated from *L. okamurai* laurinterol (**174**), debromolaurinterol (**175**), and **180–182.** Further examination of the alga collected from different localities by Suzuki and Kurosawa (1978) furnished three new compounds: neolaurinterol (**179;** mp 62°–63°, $[\alpha]_D$ −14°), isoaplysin (**184**; oil, $[\alpha]_D$ −33°), and the cyclic ether **190** (oil, $[\alpha]_D$ −25°) in addition to **174, 175,** isolaurinterol (**177**), and debromoisolaurin-terol. The latter had also been reported as a constituent of *L. intermedia* (Irie *et al.*, 1970). The structure of **179** was deduced from the comparison of spectral data with those of **174** and confirmed by synthesis. De-bromolaurinterol (**175**) was treated with *N*-bromosuccinimide in CCl₄,

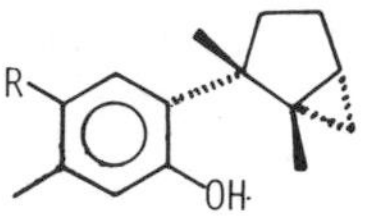

174   R = Br
175   R = H

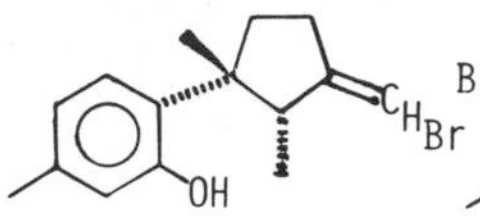

176

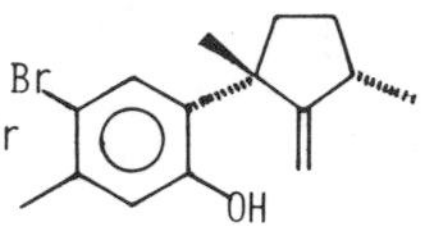

177

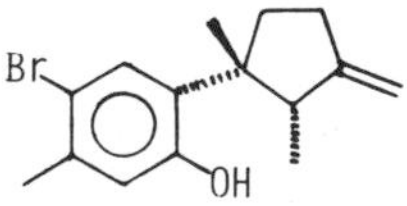

178

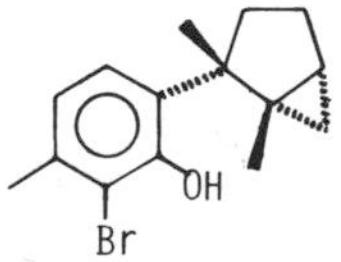

179

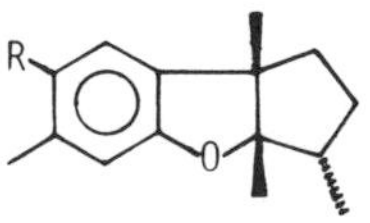

180   R = Br
181   R = H

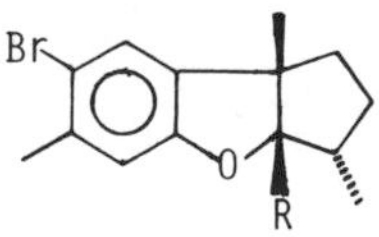

182   R = CH₂OH
183   R = CHO

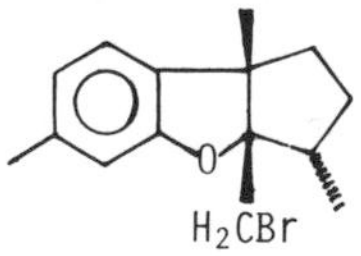

184

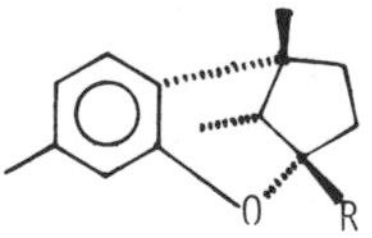

185   R = CH₂Br
186   R = CHBr₂

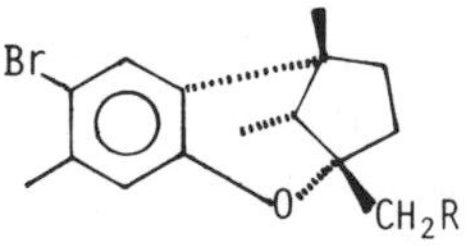

187   R = Br
188   R = H
189   R = OH

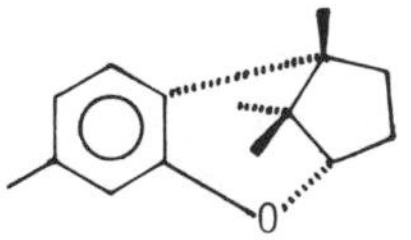

190

effecting ortho bromination to give **179.** The structures of **184** and **190** were assigned mainly by ^{1}H and ^{13}C nmr spectral analysis and by comparison with those of related compounds.

From *Laurencia filiformis* f. *heteroclada* collected on the coast of South Australia, Kazlauskas *et al.* (1976) isolated allolaurinterol (**178**; oil, $[\alpha]_D$ +22.0°), filiformin (**188**; mp 86.4°–87.3°, $[\alpha]_D$ −20.0°), and filiforminol (**189**; oil, $[\alpha]_D$ −13.7°) in addition to nonphenolic related compounds. The structures of these products were deduced on the basis of spectral data. Of the three compounds, **188** was shown to be an artifact formed by spontaneous conversion of allolaurinterol (**178**) during storage at 4°. This conversion that can be induced by the presence of a trace of acid is fully analogous to the spontaneous conversion of **176** to **185** (Irie *et al.*, 1969a) and to the transformation of laurinterol (**174**) to aplysin (**180**), which could be accomplished *in vivo* (Stallard and Faulkner, 1974b) and *in vitro* (Irie *et al.*, 1966, 1970; Suzuki *et al.*, 1969).

Ohta and Takagi (1977a) have reported the isolation of aplysinal (**183**) and four related compounds; laurinterol (**174**), debromolaurinterol (**175**), aplysin (**180**), and aplysinol (**182**), from the red alga *Marginisporum aberrans*. The new aldehyde **183** was characterized by spectral data and confirmed by its conversion to aplysinol (**182**). The authors noted that all five compounds were also found in two other species of the same family Corallinaceae, *Amphiroa zonata* and *Corallina pilulifera,* collected at the same place, where some *Laurencia* species were growing in the vicinity. Since the genus *Laurencia* (family Rhodomelaceae) and its predatory sea hares have been the exclusive sources of these metabolites, it is highly desirable to reinvestigate the Corallinaceae species in order to exclude the possibility of contamination.

VI. POLYPRENYL PHENOLS

Prenylated hydroquinones and their quinone congeners, e.g., tocopherols and ubiquinones, are widely distributed in nature. A survey by Jensen (1969) revealed that α-tocopherol is present in many species of red, brown, and green algae. Recently, a number of phenols linked to isoprenoids have been isolated from marine sources, mainly from algae and sponges. The terpenoid moiety of these compounds ranges from two to eight isoprene units and from linear to cyclic skeletons. The phenolic moieties are mainly free, but some are cyclized or alkylated hydroquinones. Some of them are even mono- and trihydric phenols. Benzoquinone congeners without a phenolic moiety have been omitted from this discussion.

A. Diprenyl Phenols

Geranylhydroquinone (**191**; oil) was isolated from the chloroform extracts of a colonial tunicate of the genus *Aplidium* (Fenical, 1976). The structure was deduced on spectral and chemical grounds. The spectral data were identical with those reported for a synthetic sample of the hydroquinone (Inouye *et al.*, 1968). Hydroquinone **191** amounted to 7% dry weight of the tunicate, and it was shown to provide protection against leukemia and tumor development in test animals.

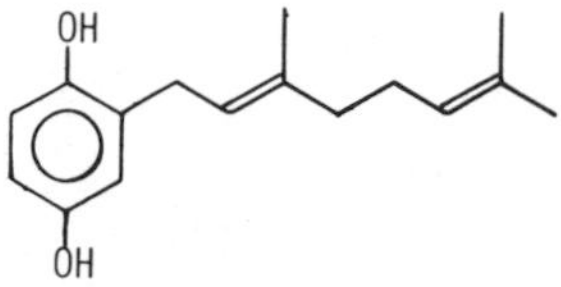

191

A number of bromo-substituted hydroquinones have been identified in lipid extracts of the green calcareous alga *Cymopolia barbata* collected in Bermuda (Högberg *et al.*, 1976). These are cymopol (**192**; mp 59–61°), cymopol monomethyl ether (**193**; oil), cymopolone (**194**; mp 79°–81°), isocymopolone (**195**; mp 55°–56°), cyclocymopol (**196**; oil), cyclocymopol monomethyl ether (**197**; glass), and cymopochromenol (**199**; oil). The structures were assigned by spectral analysis. Confirmation of structure **192** was made by condensating bromohydroquinone with geraniol in the

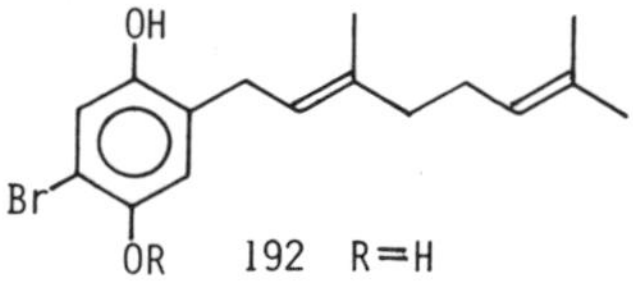

192   R=H
193   R=Me

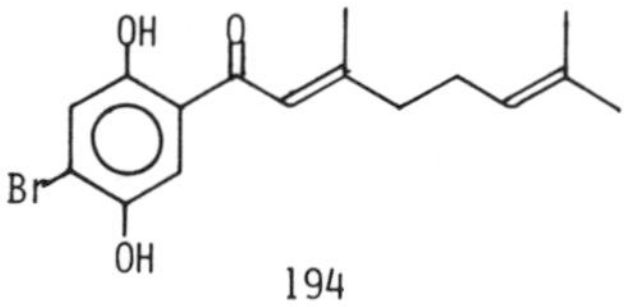

194

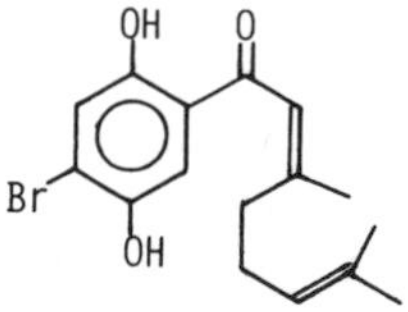

195

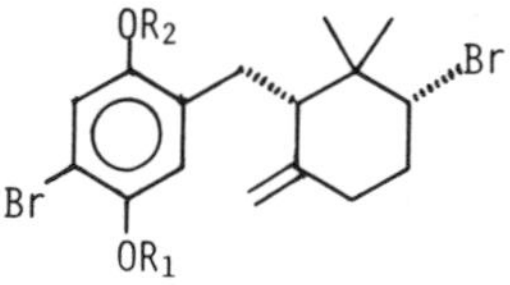

196   $R_1 = R_2 = H$
197   $R_1 = Me, R_2 = H$
198   $R_1 = Me, R_2 = Ac$

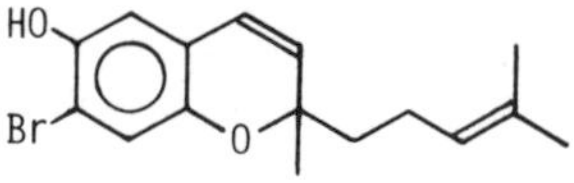

199

presence of boron trifluoride–ether complex. Single-crystal X-ray crystallographic determination of the acetate **198** confirmed structure **197** and established absolute stereochemistry. A mechanism for the stereospecific formation of cyclocymopols was postulated. It follows a route involving an attack by a bromine cation (or equivalent) on the double bond remote from the aromatic ring to form the chiral intermediate (**200**), followed by cyclization leading to **201**. Optically inactive product cymopochromenol (**199**) was suggested to be mainly an artifact derived from cymopol (**192**), which was presumably oxidized to its quinone congener and cyclized during the drying process of the alga.

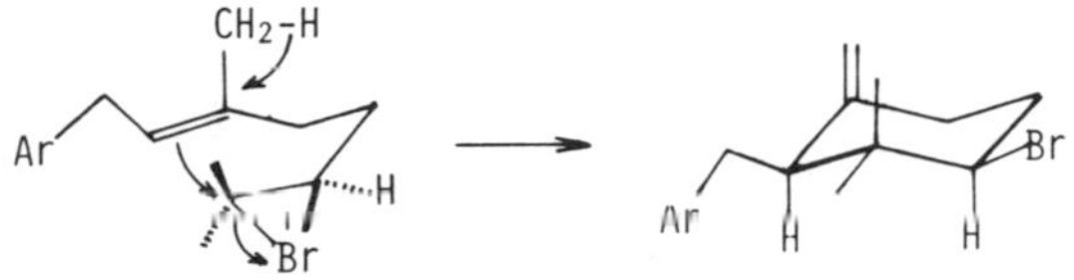

200   201

B. Triprenyl Phenols

1. Furanohydroquinone

Furan-containing terpenoids are abundant in sponges (for review, see Minale in Volume I). To date only few furanoterpenes have been isolated from soft corals (see Tursch et al., Volume II). One such compound is a furan-containing farnesylhydroquinone isolated from the soft coral *Sinularia lochmodes* (Coll et al., 1978). The hydroquinone (**202**), 2-[2,6-dimethyl-8-(4-methyl-2,5-dihydroxyphenyl)-2,6-octadienyl]-4-methylfuran (mp 88°–89°), was obtained from the dichloromethane extract of the freeze-dried soft coral and was assigned a structure based largely on ¹H and ¹³C nmr data and by comparison with those of the corresponding quinone.

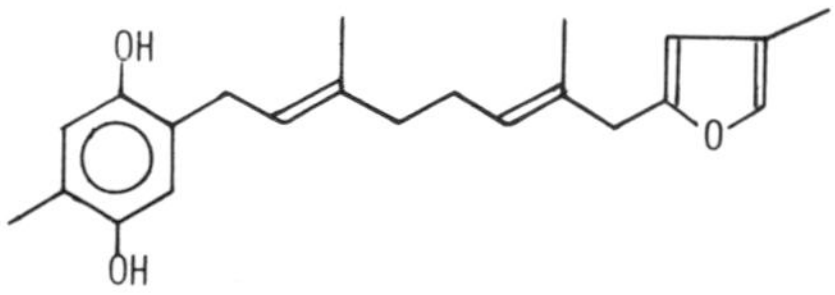

202

2. Zonarol and Related Compounds

These are drimane-type sesquiterpene-substituted hydroquinones and their derivatives from a brown alga and from sponges. The algal products

(**203–207**) were reviewed by Martin and Darias (1978) in Volume I. The only addition to these is yahazunol (**208**; mp 127°–129°, $[\alpha]_D$ −12°), recently isolated from the alga *Dictyopteris undulata* (Ochi *et al.*, 1979). The structure of **208** was deduced on spectral grounds and confirmed by synthesis from zonarol (**203**). Epoxide **210** was prepared by treatment of the dimethyl ether of **203** with *m*-chloroperbenzoic acid and reduced by reaction with lithium aluminum hydride to furnish a tertiary alcohol (**209**). The alcohol was identical with the dimethyl ether derived from **208**.

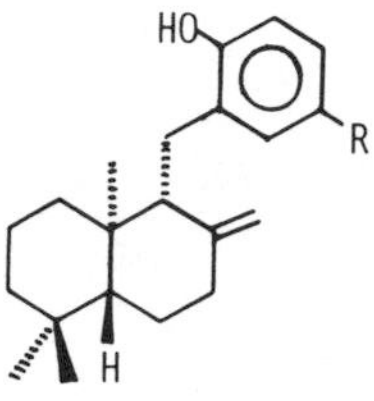

203   R = OH
204   R = COOH   205 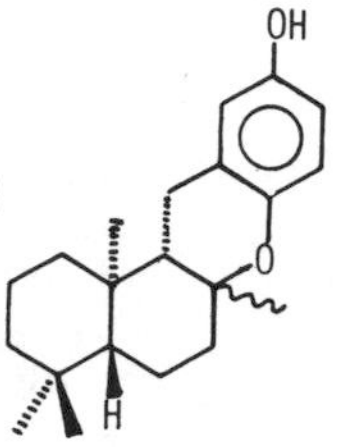   206

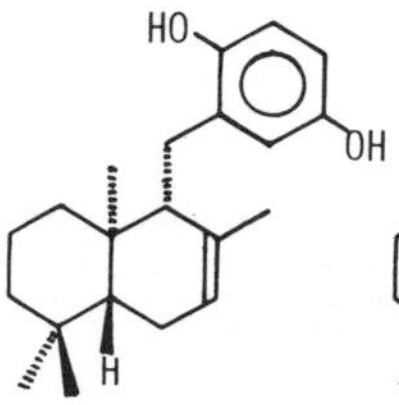

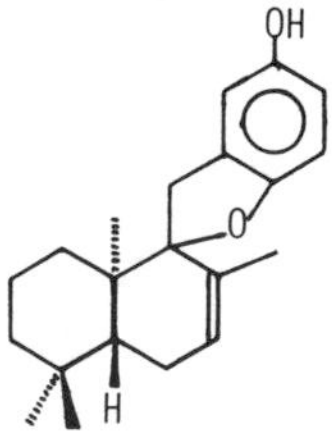

207 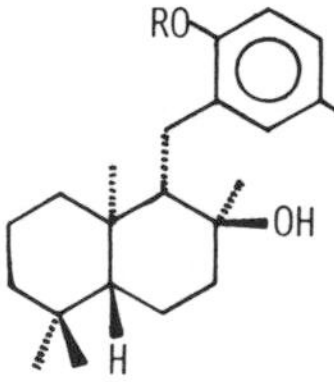   208   R = H   210
             209   R = Me

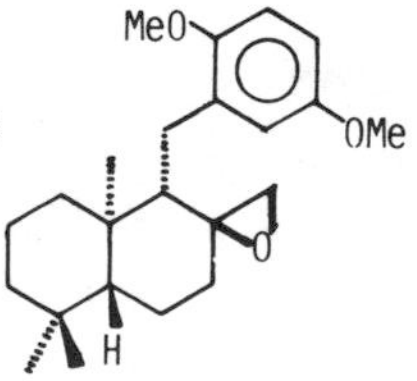

In addition to **208** the same Japanese group isolated from the alga crystalline zonarol (**203**; mp 173.5°–174.5°), isozonarol (**205**; mp 116°–117°), and zonaroic acid (**204**; mp 108–109°) which had previously been described as noncrystalline gums. The co-occurrence of the benzoic acid (**204**) and the hydroquinone (**203**) in the alga was suggested as evidence that *p*-hydroxybenzoic acid is the precursor of the phenolic moiety as in ubiquinone biogenesis (Cimino *et al.*, 1975d).

The sponges *Dysidea avara* and *D. pallescens* have provided closely related compounds, avarol (**211**) and *ent*-chromazonarol (**212**), respectively (Minale *et al.*, 1974; De Rosa *et al.*, 1976; Cimino *et al.*, 1975c). Phenol **212** is the optical antipode of **206**. Structural elucidation and stereochemistry of these compounds have been fully reviewed by Minale (1978) in Volume I.

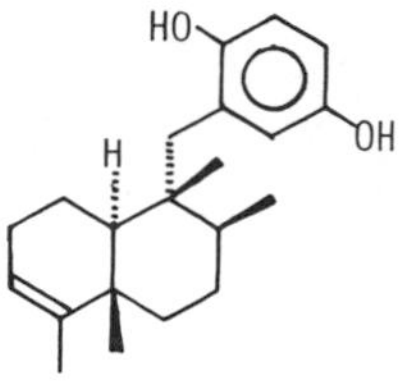

211 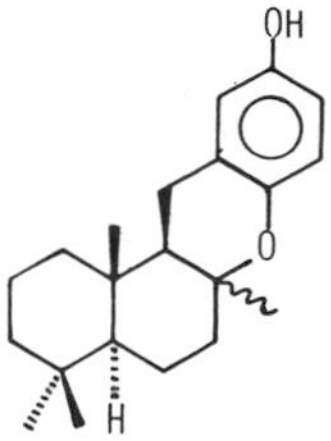   212

3. Paniceins

A series of farnesyl-derived phenols, the paniceins (**213–217**), were isolated by Cimino *et al.* (1973) from the sponge *Halichondria panicea*. For details a review by Minale (1978) should be consulted.

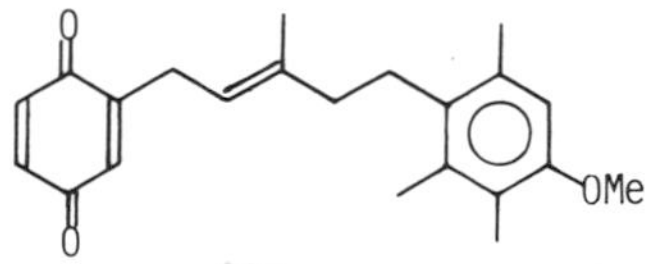

213 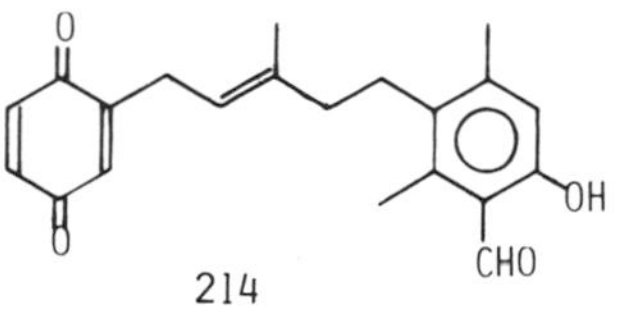   214

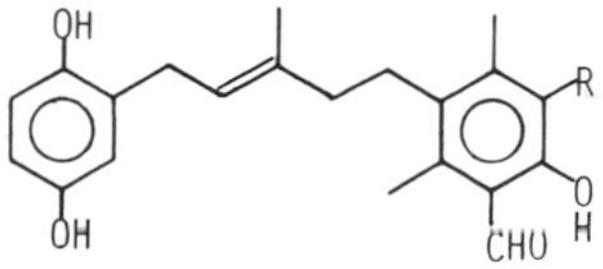

215   R = H
216   R = OH   217

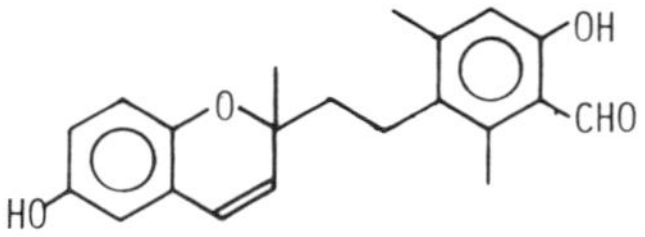

C. Tetraprenyl Phenols

A number of tetraprenyl hydroquinones and derivatives have been described as constituents of a sponge (Cimino *et al.*, 1972b) and several species of brown algae (González *et al.*, 1971, 1973a,b,c, 1974; Kato *et al.*, 1975a, b; Kikuchi *et al.*, 1975; Gregson *et al.*, 1977; Gerwick *et al.*, 1979). Many of these compounds (**218–223**) have been discussed in preceding volumes of this series (see Minale in Volume I and Fenical in Volume II). Four additional compounds (**224–226, 229**) enter the present discussion.

Two functional derivatives (**224, 225**) of δ-tocotrienol (**220**) have been isolated from the brown alga *Cystophora torulosa* together with the polyketide-derived resorcinols and phloroglucinol (*vide supra*) (Gregson *et al.*, 1977). The structures of these compounds were assigned on the basis of spectral data and by comparison with **220**. The position of the

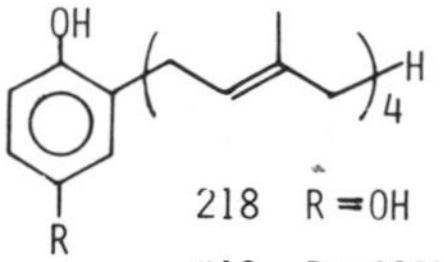

218 R = OH
219 R = COOH

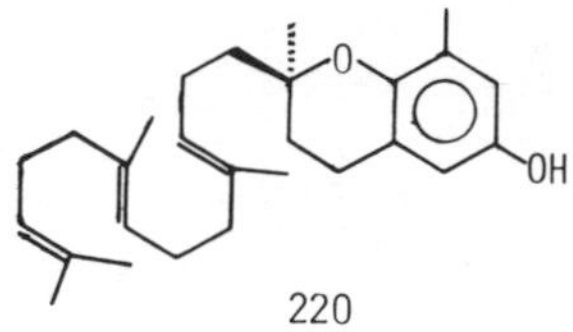

220

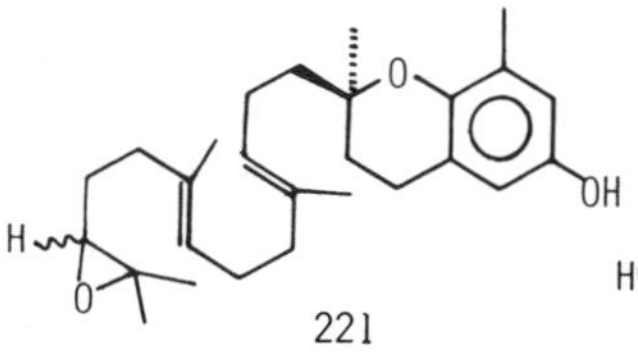

221

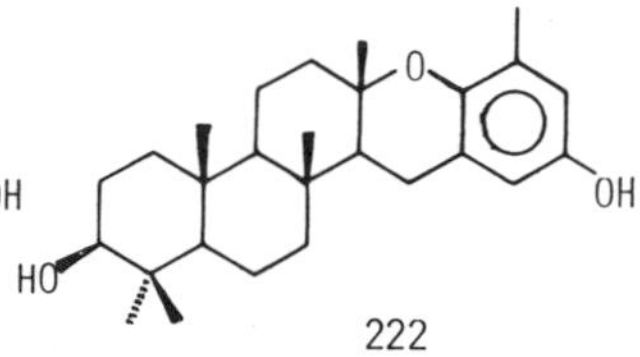

222

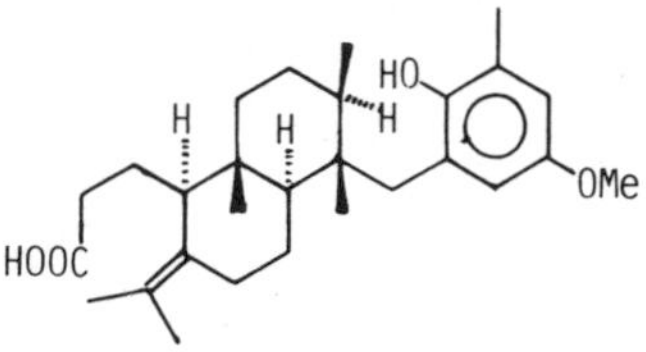

223

additional methyl group in **225** has not been established. Unlike **220** and **221**, which are known to be effective inducers for settling of the swimming larvae of the hydrozoan *Coryne uchidai* (Kato *et al.*, 1975a), **224** and **225** appeared to have no such effect on hydroid larvae, since several collections of the alga from Australia were devoid of hydroid symbionts.

The alga *Sargassum tortile,* which yielded phenols **220** and **221**, contains another related compound, sargatriol (**226**; oil, $[\alpha]_D$ +16°) (Kikuchi *et al.*, 1975). The structure of **226** was assigned on spectral and chemical

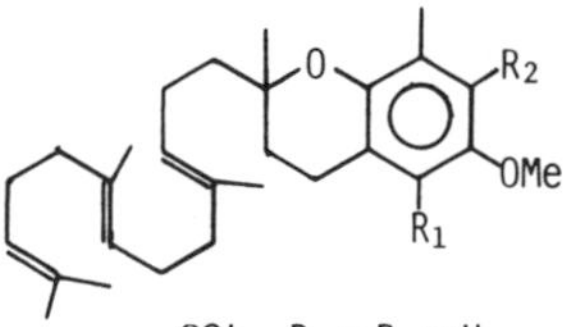

224 $R_1 = R_2 = H$
225 $R_1 = Me,$ $R_2 = H$ or
     $R_1 = H,$ $R_2 = Me$

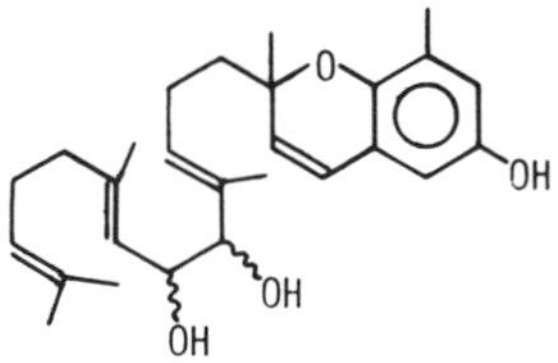

226

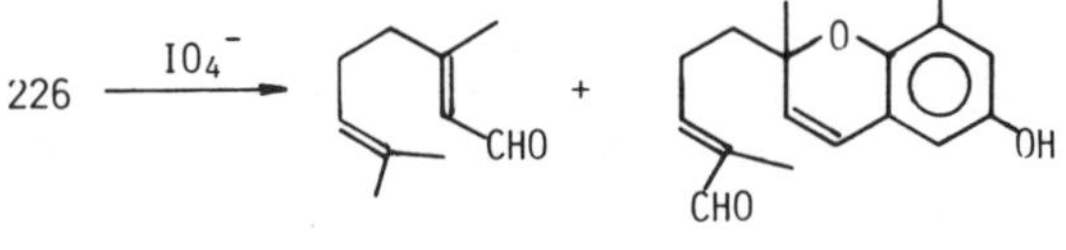

$$226 \xrightarrow{IO_4^-}$$

227   228

grounds. Periodate oxidation of **226** furnished two conjugated aldehydes, **227** and **228,** which were identified by spectral analysis. Aldehyde **227** was proved identical with citral-a (Ohtsuru *et al.*, 1967). Catalytic hydrogenation of the triacetate of **226** removed two vicinal acetoxy groups and yielded a compound identical with one derived from **220.**

In a recent communication Gerwick *et al.* (1979) reported isolation and structures of novel ichthyotoxic constituents of the tropical brown alga *Stypopodium zonale* which, when placed in an aquarium, turns the water a rust color, and is extremely toxic to the reef-dwelling herbivorous damsel fish *Eupomacentrus leucostictus*. Silica gel column chromatography of the crude chloroform/methanol extract of *S. zonale* yielded stypoldione [**229**; mp 170° (dec), $[\alpha]_D$ −65.1°] as the major toxin. Pure **229,** however, was less toxic than the crude extract, from which more toxic stypotriol (**230**; $[\alpha]_D$ −10.0°) could be isolated by rapid chromatography. Pure stypotriol was rapidly air-oxidized to the quinone **229,** but it was stable in the crude extract. These toxins were characterized by spectroscopic methods, and the structure of **229** was determined by X-ray crystallography. Absolute configuration has not been established.

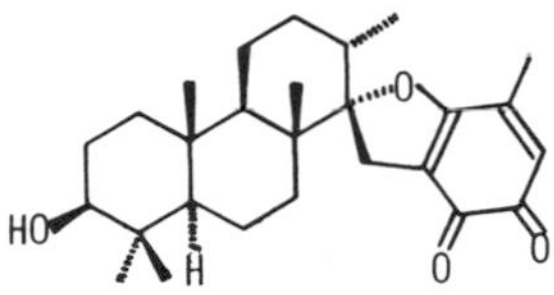

229 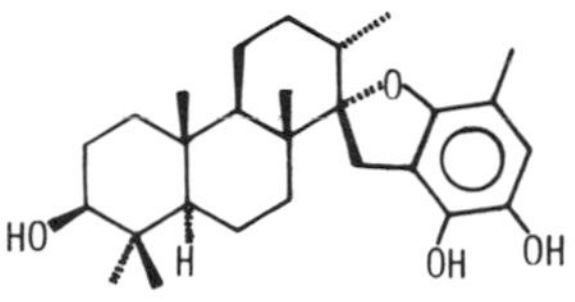   230

In another recent paper, Kusumi *et al.* (1979) described isolation of sargachromenol (**231**) along with two new plastoquinones from the brown alga *Sargassum serratifolium*. But the phenol **231** appears to be an artifact formed during the extraction procedure as demonstrated by ready conversion of one of the quinones (**232**).

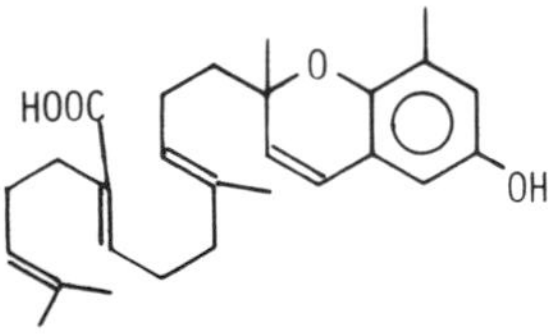

231 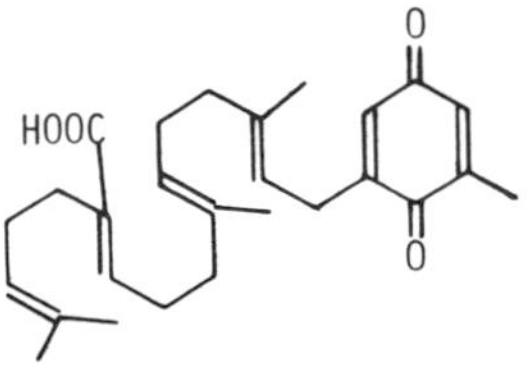   232

D. Other Polyprenyl Phenols

Minale's group in Italy has isolated desidein (**233**), a pentacyclic sesterterpene condensed with hydroxyhydroquinone, from the sponge *Dysidea*

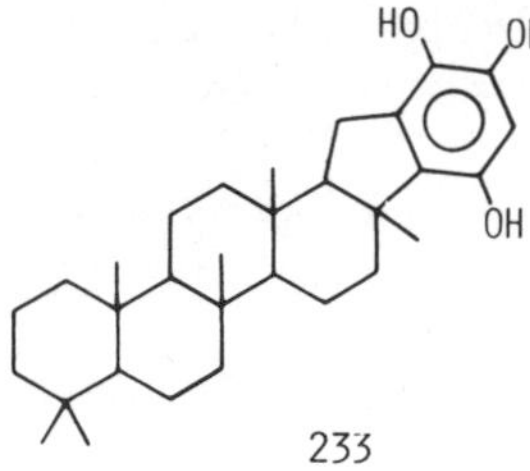

233

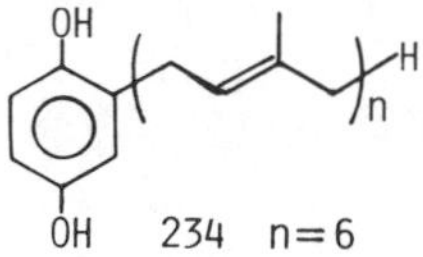

234　n=6
235　n=7
236　n=8

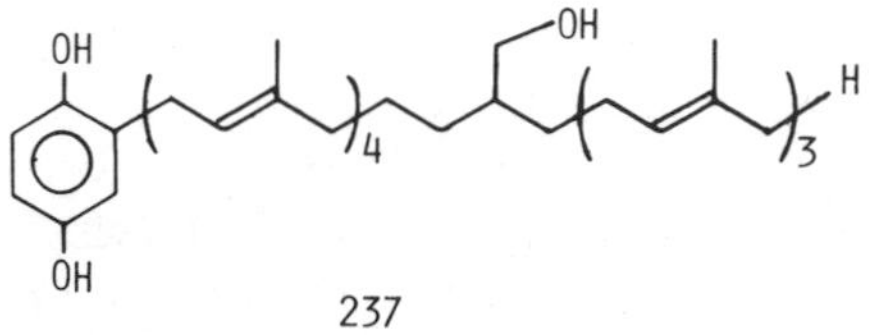

237

pallescens (Cimino *et al.*, 1975a). The same group had previously isolated a series of linear polyprenyl hydroquinones (**234–237**) from another sponge, *Ircinia spinosula,* along with the corresponding benzoquinones (Cimino *et al.*, 1972a). All these compounds have been discussed by Minale in Volume I.

APPENDIX: PHYLETIC DISTRIBUTION OF PHENOLIC SUBSTANCES

TABLE A1

Distribution of Phenolic Substances in Marine Plants

Species	Phenolic substances	References
Phylum Cyanophyta		
Order Holmogonales		
Family Oscillatoriaceae		
Lyngbya majuscula	**68, 69**	Marner *et al.* (1977)
L. gracilis	**160**	Mynderse *et al.* (1977)
Oscillatoria nigroviridis	**160, 163–166**	Mynderse and Moore (1978)
Schizothrix calciola		
Family Rivulariaceae		
Calothrix brevissima	**21–23**	Pedersén and Dasilva (1973)
Phylum Rhodophyta		
Order Gigartinales		
Family Rhodophyllaceae		
Cystoclonium purpureum	**32**	Pedersén *et al.* (1974)
Family Phyllophoraceae		
Phyllophora nervosa	**45**	Güven *et al.* (1970)

TABLE A1 (*Continued*)

Species	Phenolic substances	References
Order Cryptonemiales		
Family Corallinaceae		
Corallina officinalis	33	Pedersén *et al.* (1974)
C. pilulifera	174, 177, 180, 182, 183	Ohta and Takagi (1977a)
Amphiroa zonata	174, 177, 180 182, 183	Ohta and Takagi (1977a)
Marginisporum aberrans	14, 19, 174, 177, 180, 182, 183	Ohta and Takagi (1977a,b)
Order Ceramiales		
Family Rhodomelaceae		
Polysiphonia lanosa	21, 31–35, 39, 40	Glombitza and Stoffelen (1972) Hodgkin *et al.* (1966) Glombitza *et al.* (1974)
P. nigrescens	21, 31–33, 35, 39, 40, 96	Glombitza *et al.* (1974) Pedersén *et al.* (1974) Pedersén (1978)
P. brodiaei	20, 21, 32, 33, 96	Glombitza *et al.* (1974) Pedersén *et al.* (1974) Lundgren *et al.* (1979)
P. fruticulosa	20, 32, 33	Glombitza *et al.* (1974)
P. nigra	20, 32, 33	Glombitza *et al.* (1974)
P. urceolata	20, 31–33, 35, 99	Pedersén *et al.* (1974) Kurata *et al.* (1976)
P. elongata	31–33	Glombitza *et al.* (1974)
P. violacea	32, 33	Glombitza *et al.* (1974)
P. thuyoides	32, 33	Glombitza *et al.* (1974)
P. morrowii	31	Saito and Ando (1955)
Odonthalia dentata	21, 31–33	Craigie and Gruenig (1967) Pedersén *et al.* (1974)
O. corymbifera	33, 40	Kurata *et al.* (1973)
Rhodomela subfusca	32–34, 40	Kurata and Amiya (1975) Glombitza *et al.* (1974)
R. confervoides	21, 32, 33, 96	*Craigie and Gruenig (1967)* *Pedersén et al.* (1974) Pedersén (1978) Kurata and Amiya (1977)
R. larix	32, 33, 40, 96. 97	Katsui *et al.* (1967) Weinstein *et al.* (1975) Kurata and Amiya (1977)
Halopytis pinastroides (syn. *H. incurvus*)	20, 32, 35, 61, 77, 105	Glombitza *et al.* (1974) Chantraine *et al.* (1973) Combaut *et al.* (1978a,b)
Rytiphlea tinctoria	33, 103, 104	Chevolot-Magueur *et al.* (1976)
Laurencia intermedia	174, 175, 177	Irie *et al.* (1966, 1970)
L. nipponica	174, 176	Irie *et al.* (1969a)

TABLE A1 (*Continued*)

Species	Phenolic substances	References
L. okamurai	**174, 175, 184,**	Irie *et al.* (1969b)
	179–182, 190	Suzuki and Kurosawa (1978)
L. nidifica	**174, 180**	Waraszkiewicz and Erickson (1974)
L. pacifica	**174**	Sims *et al.* (1971)
L. filiformis	**178, 188, 189**	Kazlauskas *et al.* (1976)
L. glandulifera	**185–187**	Suzuki and Kurosawa (1976)
Family Ceramiaceae		
Antithamnion plumula	**33**	Pedersén *et al.* (1974)
Ceramium rubrum	**33**	Pedersén *et al.* (1974)
Family Delesseriaceae		
Martensia fragilis	**26**	Moore (1977)
Family Dasyaceae		
Dasya pedicellata	**18, 19**	Fenical and McConnell (1976)
var. *stanfordiana*		
Phylum Phaeophyta		
Order Fucales		
Family Fucaceae		
Fucus vesiculosus	**5, 21, 33, 132,**	Pedersén and Fries (1975)
	133–135, 139–143	Ragan and Craigie (1976)
		Glombitza *et al.* (1975a, 1977c)
		Craigie *et al.* (1977)
F. serratus	**5**	Glombitza *et al.* (1973)
F. spiralis	**5**	Glombitza *et al.* (1973)
Ascophyllum nodosum	**153**	Jensen and Ragan (1978)
Family Himanthaliaceae		
Himanthalia elongata	**5**	Glombitza *et al.* (1973)
Family Cystoseiraceae		
Cystoseira tamariscifolia	**5, 136, 148**	Glombitza *et al.* (1973, 1975b)
C. baccata	**5, 132, 145,**	Glombitza *et al.* (1973,
	146, 148	1978b)
C. granulata	**5**	Glombitza *et al.* (1973)
C. myriophylloides	**5**	Glombitza *et al.* (1973)
Bifurcaria bifurcata	**5, 132, 133, 136,**	Glombitza and Rösener (1974)
	148, 149, 151, 152	Glombitza *et al.* (1973, 1976b)
Halidrys siliquosa	**5, 149, 151, 152**	Glombitza and Sattler (1973)
		Sattler *et al.* (1977)
		Glombitza *et al.* (1973)
		Glombitza (1977)
Cystophora torulosa	**154–158, 224, 225**	Gregson *et al.* (1977)
Family Sargassaceae		
Sargassum muticum	**5, 136, 148–150**	Glombitza *et al.* (1978a)
S. tortile	**220, 221, 226**	Kato *et al.* (1975b)
		Kikuchi *et al.* (1975)

TABLE A1 (*Continued*)

Species	Phenolic substances	References
S. serratifolium	**231**	Kusumi *et al.* (1979)
Order Dictyotales		
Family Dictyotaceae		
Dictyota dichotoma	**5, 132, 136**	Glombitza *et al.* (1973, 1977a)
Taonia atomaria	**222, 223**	González *et al.* (1973b, 1974)
Dictyopteris undulata	**203–208**	Fenical *et al.* (1973)
(*D. zonarioides*)		Fenical and McConnell (1975)
		Cimino *et al.* (1975d)
		Ochi *et al.* (1979)
Stypopodium zonale	**230**	Gerwick *et al.* (1979)
Order Laminariales		
Family Laminariaceae		
Laminaria ochroleuca	**5, 136–138, 147**	Glombitza *et al.* (1973, 1976a, 1977b)
		Glombitza (1977)
Saccorhiza polyschides	**5**	Glombitza *et al.* (1973)
Family Chordaceae		
Chorda filum	**5**	Glombitza *et al.* (1973)
Order Sphacelariales		
Family Cladostephaceae		
Cladostephus spongiosus	**5**	Glombitza *et al.* (1973)
C. verticillatus	**5**	Glombitza *et al.* (1973)
Phylum Chlorophyta		
Order Ulvales		
Family Monostromaceae		
Monstroma fuscum	**44**	Tocher and Craigie (1966)
Order Siphonales		
Family Dasycladaceae		
Cymopolia barbata	**192–197, 199**	Högberg *et al.* (1976)
Order Charales		
Family Characeae		
Nitella hookeri	**91, 92**	Markham and Porter (1969)
Phylum Spermatophyta		
Posidonia oceanica	**27–29, 83**	Cariello *et al.* (1979)
Cymodocea nodosa	**82, 83, 89, 90**	Cariello *et al.* (1979)

TABLE A2

Distribution of Phenolic Substances in Marine Animals

Species	Phenolic substances	References
Phylum Porifera		
Order Keratosa		
Verongia archeri	**55**	Stempien *et al.* (1973)
V. aurea	**59, 60**	Krejcarek *et al.* (1975)

TABLE A2 (*Continued*)

Species	Phenolic substances	References
V. lacunosa	48	Borders *et al.* (1974)
Psammoposilla purpurea	56	Chang and Weinheimer (1977)
Dysidea avara	211	Minale *et al.* (1974)
D. herbacea	117–122	Sharma *et al.* (1970)
		Sharma and Vig (1972)
D. pallescens	212, 233	Cimino *et al.* (1975a,c)
Ircinia muscarum	218, 219	Cimino *et al.* (1972b)
I. spinosula	234–237	Cimino *et al.* (1972a)
Spongia officinalis	63–66	Low (1951)
obliqua		Ackerman and Müller (1941)
Order Halichondrida		
Halichondria panicea	213–217	Cimino *et al.* (1973)
Hymeniacidon sanguinea	79	Cimino *et al.* (1975b)
Order Axinellida		
Axinella polypoides	15	Cimino *et al.* (1974)
Order Hadromerida		
Cliona celata	49, 75, 76	Andersen (1978)
		Stonard and Andersen (1979)
Anthosigmella varians	58	Schmitz *et al.* (1977)
Spheciospongia vesparia	86	Bergmann and McAleer (1951)
Phylum Coelenterata		
Order Gorgonacea		
Gorgonia cavolinii	63	Drechsel (1907)
Primnoa lepudifera	65	Mörner (1913)
Subergorgia hicksoni	172, 173	Kashman (1979)
Psudopterogorgia rigida	169, 170	McEnroe and Fenical (1978)
Order Zoantharia		
Palythoa sp.	47	Sheikh (1969)
Order Alcyonacea		
Sinularia lochmodes	202	Coll *et al.* (1978)
Phylum Mollusca		
Class Gastropoda		
Subclass Prosobranchia		
Buccinum undatum	67	Hunt and Breuer (1971)
Subclass Opistobranchia		
Stylocheilus longicauda	159–162	Kato and Scheuer (1974)
Aplysia kurodai	180–181	Yamamura and Hirata (1963)
A. californica	174, 180, 181	Stallard and Faulkner (1974a)
A. brasiliana	167	Kinnel *et al.* (1977)
Navanax innermis	95	Sleeper and Fenical (1977)
Class Cephalopoda		
Subclass Metacephalopoda		
Octopus vulgaris	46, 70–72	Erspamer (1952)
		Ito *et al.* (1976)
Phylum Annelida		
Class Polychaeta		
Order Sedentaria		
Thelepus setosus	21, 24, 109–111	Higa and Scheuer (1975a)

TABLE A2 (*Continued*)

Species	Phenolic substances	References
Lanice conchilega	**8, 17**	Weber and Ernst (1978)
Phylum Phoronida		
Phoronopsis viridis	**6, 8**	Sheikh and Djerassi (1975)
Phylum Hemichordata		
Class Enteropneusta		
Ptychodera flava	**8, 11, 128**	Higa *et al.* (1979)
P. flava Laysanica	**8, 11, 12, 124,**	Higa and Scheuer (1977)
	126, 128	
Glossobalanus sp.	**3, 8–10**	Higa *et al.* (1980)
Balanoglossus carnosus	**6–8, 11**	Higa *et al.* (1980)
B. misakiensis	**6, 10**	Higa *et al.* (1980)
B. biminiensis	**6**	Ashworth and Cormier (1967)
Phylum Protochordata		
Aplidium sp.	**191**	Fenical (1976)

REFERENCES

Ackermann, D., and Müller, E. (1941). *Hoppe-Seyler's Z. Physiol. Chem.* **269,** 146.

Andersen, R. J. (1978). *Tetrahedron Lett.* p. 2541.

Andersen, R. J., and Stonard, R. J. (1979). *Can. J. Chem.* **57,** 2325.

Andersen, R. J., Wolfe, M. S., and Faulkner, D. J. (1974). *Mar. Biol.* **27,** 281.

Ashworth, R. B., and Cormier, M. J. (1967). *Science* **155,** 1558.

Baker, W., and Enderby, T. (1940). *J. Chem. Soc.,* p. 1094.

Barrington, E. J. W., and Thorpe, A. (1963). *Gen. Comp. Endocrinol.* **3,** 166.

Bergmann, W., and McAleer, W. J. (1951). *J. Am. Chem. Soc.* **73,** 4946.

Birch, A. J., Massy-Westropp, R. A., Richards, R. W., and Smith, II. (1958). *J. Chem. Soc.* p. 360.

Borders, D. B., Morton, G. O., and Wetzel, E. R. (1974). *Tetrahedron Lett.* p. 2709.

Burkholder, P. R., Pfister, R. M., and Leitz, F. H. (1966). *Appl. Microbiol.* **14,** 649.

Cariello, L., Zanetti, L., and De Stefano, S. (1979). *Comp. Biochem. Physiol.* **62B,** 159.

Chang, C. W. J., and Weinheimer, A. J. (1977). *Tetrahedron Lett.,* p. 4005.

Chantraine, J.-M., Combaut, G., and Teste, J. (1973). *Phytochemistry* **12,** 1793.

Charles, C., Braekman, J. C., Daloze, D., Tursch, B., and Karlsson, R. (1978). *Tetrahedron Lett.,* p. 1519.

Chevolot-Magueur, A.-M., Cave, A., Potier, P., Teste, J., Chiaroni, A., and Riche, C. (1976). *Phytochemistry* **15,** 767.

Cimino, G., De Stefano, S., and Minale, L. (1972a). *Tetrahedron* **28,** 1315.

Cimino, G., De Stefano, S., and Minale, L. (1972b). *Experientia* **28,** 1401.

Cimino, G., De Stefano, S., and Minale, L. (1973). *Tetrahedron* **29,** 2565.

Cimino, G., De Stefano, S., and Minale, L. (1974). *Comp. Biochem. Physiol.* **47B,** 895.

Cimino, G., De Luca, P., De Stefano, S., and Minale, L. (1975a). *Tetrahedron* **31,** 271.

Cimino, G., De Stefano, S., and Minale, L. (1975b). *Experientia* **31,** 756.

Cimino, G., De Stefano, S., and Minale, L. (1975c). *Experientia* **31,** 1117.

Cimino, G., De Stefano, S., Fenical, W., Minale, L., and Sims, J. J. (1975d). *Experientia* **31,** 1250.

Coll, J. C., Liyanage, N., Stokie, G. J., Altena, I. V., Nemorin, J. N. E., Sternhell, S., and Kazlauskas, R. (1978). *Aust. J. Chem.* **31,** 157.

Combaut, G., Chantraine, J.-M., Teste, J., and Glombitza, K.-W. (1978a). *Phytochemistry* **17,** 1791.

Combaut, G., Chantraine, J.-M., Teste, J., Soulier, J., and Glombitza, K.-W. (1978b). *Tetrahedron Lett.,* p. 1699.

Cookson, R. C., Halton, B., and Stevens, I. D. R. (1968). *J. Chem. Soc. (B)* p. 767.

Craigie, J. S., and Gruenig, D. E. (1967). *Science* **157,** 1058.

Craigie, J. S., and McLachlan, J. (1964). *Can. J. Bot.* **42,** 23.

Craigie, J. S., McInnes, A. G., Ragan, M. A., and Walter, J. A. (1977). *Can. J. Chem.* **55,** 1575.

Crato, E. (1892). *Ber. Dtsch. Bot. Ges.* **10,** 295.

De Rosa, S., Minale, L., Riccio, R., Sodano, G. (1976). *J. Chem. Soc. Perkin I* p. 1408.

Drechsel, E. (1907). *Z. Biol.* **33,** 85.

Erspamer, V. (1952). *Nature (London)* **169,** 375.

Fattorusso, E., Minale, L., and Sodano, G. (1972). *J. Chem. Soc. Perkin I* p. 16.

Faulkner, D. J. (1978). *In* "Topics in Antibiotic Chemistry" (P. G. Sammes, ed.), Vol. 2, p. 13. Ellis Horwood, Chichester.

Faulkner, D. J., and Andersen, R. J. (1974). *In* "The Sea" (E. D. Goldberg, ed.), Vol. 5, p. 679. Wiley (Interscience), New York.

Fenical, W. (1975). *J. Phycol.* **11,** 245.

Fenical, W. (1976). *In* "Food-Drugs from the Sea Proceedings 1974" (H. H. Webber and G. D. Ruggieri, eds.), p. 388. Marine Technology Society, Washington, D.C.

Fenical, W., and McConnell, O. (1975). *Experientia* **31,** 1004.

Fenical, W., and McConnell, O. (1976). *Phytochemistry* **15,** 435.

Fenical, W., Sims, J. J., Squatrito, D., Wing, R. M., and Radlick, P. (1973). *J. Org. Chem.* **38,** 2383.

Fries, L. (1973). *Experientia* **29,** 1436.

Fukuzawa, A., and Kurosawa, E. (1979). *Tetrahedron Lett.,* p. 2797.

Geissmann, T. A. (1967). *In* "Biogenesis of Natural Compounds" (P. Bernfeld, ed.), 2nd ed., p. 563. Pergamon, Oxford.

Geissman, T. A., and Crout, D. H. G. (1969). "Organic Chemistry of Secondary Plant Metabolism." Freeman, San Francisco, California.

Gerwick, W. H., Fenical, W., Fritsch, N., and Clardy, J. (1979). *Tetrahedron Lett.,* p. 145.

Glombitza, K.-W. (1977). *In* "Marine Natural Products Chemistry" (D. J. Faulkner and W. H. Fenical, eds.), p. 191. Plenum Press, New York.

Glombitza, K.-W. (1979). *In* "Marine Algae in Pharmaceutical Science" (H. A. Hoppe, T. Levring, and Y. Tanaka, eds.), pp. 303–342. Walter de Gruyter, Berlin.

Glombitza, K.-W., and Rösener, H.-U. (1974). *Phytochemistry* **13,** 1245.

Glombitza, K.-W., and Sattler, E. (1973). *Tetrahedron Lett.,* p. 4277.

Glombitza, K.-W., and Stoffelen, H. (1972). *Planta Med.* **22,** 391.

Glombitza, K.-W., Rösener, H.-U., Vilter, H., and Rauwald, W. (1973). *Planta Med.* **24,** 301.

Glombitza, K.-W., Stoffelen, H., Murawski, U., Bielaczek, J., and Egg, H. (1974). *Planta Med.* **25,** 105.

Glombitza, K.-W., Rauwald, H.-W., and Eckhardt, G. (1975a). *Phytochemistry* **14,** 1403.

Glombitza, K.-W., Rösener, H.-U., and Müller, D. (1975b). *Phytochemistry* **14,** 1115.

Glombitza, K.-W., Koch, M., and Eckhardt, G. (1976a). *Phytochemistry* **15,** 1082.

Glombitza, K.-W., Rösener, H.-U., and Koch, M. (1976b). *Phytochemistry* **15,** 1279.

Glombitza, K.-W., Geisler, C., and Eckhardt, G. (1977a). *Phytochemistry* **16,** 2035.

Glombitza, K.-W., Koch, M., and Eckhardt, G. (1977b). *Phytochemistry* **16,** 796.

Glombitza, K.-W., Rauwald, H.-W., and Eckhardt, G. (1977c). *Planta Med.* **32**, 33.
Glombitza, K.-W., Forster, M., and Eckhardt, G. (1978a). *Phytochemistry* **17**, 579.
Glombitza, K.-W., Wiedenfeld, G., and Eckhardt, G. (1978b). *Arch. Pharm.* **311**, 393.
González, A. G., Darias, J., and Martín, J. D. (1971). *Tetrahedron Lett.* p. 2729.
González, A. G., Alvarez, M. A., Darias, J., and Martín, J. D. (1973a). *J. Chem. Soc. Perkin 1*, p. 2637.
González, A. G., Darias, J., Martín, J. D., and Pascual, C. (1973b). *Tetrahedron* **29**, 1605.
González, A. G., Martín, J. D., and Rodriguez, M. L. (1973c). *Tetrahedron Lett.* p. 3657.
González, A. G., Darias, J., Martín, J. D., and Norte, M. (1974). *Tetrahedron Lett.* p. 3951.
Goreau, T. F., and Hartman, W. D. (1963). *In* "Mechanisms of Hard Tissue Destruction" (R. R. Sognnaes, ed.), Vol. 75, p. 25. American Association Advancement Science.
Gregson, R. P., Kazlauskas, R., Murphy, P. T., and Wells, R. J. (1977). *Aust. J. Chem.* **30**, 2527.
Grossert, J. S. (1972). *Chem. Soc. Rev.* **1**, 1.
Grove, J. F. (1963). *Q. Rev. Chem. Soc.* **17**, 1.
Güven, K. C., Bora, A., and Sunam, G. (1970). *Phytochemistry* **9**, 1893.
Hanessian, S., and Kaltenbronn, J. S. (1966). *J. Am. Chem. Soc.* **88**, 4509.
Higa, T., and Scheuer, P. J. (1974). *J. Am. Chem. Soc.* **96**, 2246.
Higa, T., and Scheuer, P. J. (1975a). *Tetrahedron* **31**, 2379.
Higa, T., and Scheuer, P. J. (1975b). *Naturwissenshaften* **62**, 395.
Higa, T., and Scheuer, P. J. (1977). *In* "Marine Natural Products Chemistry" (D. J. Faulkner and W. H. Fenical, eds.), p. 35. Plenum Press, New York.
Higa, T., Fujiyama, T., and Scheuer, P. J. (1980). *Comp. Biochem. Physiol. B* **65B**, 525.
Hodgkin, J. H., Craigie, J. S., and McInnes, A. G. (1966). *Can. J. Chem.* **44**, 74.
Hofheinz, W., and Oberhänsli, W. E. (1976). *Helv. Chim. Acta* **60**, 660.
Högberg, H.-E., Thomson, R. H., and King, T. J. (1976). *J. Chem. Soc. Perkin I* p. 1696.
Hunt, S., and Breuer, S. W. (1971). *Biochem. Biophys. Acta* **252**, 401.
Inouye, H., Tokura, K., and Tobita, S. (1968). *Chem. Ber.* **101**, 4057.
Irie, T., Suzuki, M., Kurosawa, E., and Masamune, T. (1966). *Tetrahedron Lett.* p. 1837.
Irie, T., Fukuzawa, A., Izawa, M., and Kurosawa, E. (1969a). *Tetrahedron Lett.* p. 1343.
Irie, T., Suzuki, M., and Hayakawa, Y. (1969b). *Bull. Chem. Soc. Jpn.* **42**, 843.
Irie, T., Suzuki, M., Kurosawa, E., and Masamune, T. (1970). *Tetrahedron* **26**, 3271.
Ito, S., Nardi, G., and Prota, G. (1976). *J. Chem. Soc. Chem. Commun.*, p. 1042.
Jensen, A. (1969). *J. Sci. Food Agric.* **20**, 449.
Jensen, A., and Ragan, M. A. (1978). *Tetrahedron Lett.*, p. 847.
Kashman, Y. (1979). *Tetrahedron* **35**, 263.
Katayama, T. (1961). *Nippon Suisan Gakkaishi* **27**, 75.
Kato, Y., and Scheuer, P. J. (1974). *J. Am. Chem. Soc.* **96**, 2245.
Kato, Y., and Scheuer, P. J. (1975). *Pure Appl. Chem.* **41**, 1.
Kato, Y., and Scheuer, P. J. (1976). *Pure Appl. Chem.* **48**, 29.
Kato, T., Kummaniering, S., Ichinose, I., Kitahara, Y., Kakinuma, Y., and Kato, Y. (1975a). *Chem. Lett.* p. 335.
Kato, T. *et al.* (1975b). *Experientia* **31**, 433.
Katsui, N., Suzuki, Y., Kitamura, S., and Irie, T. (1967). *Tetrahedron* **23**, 1185.
Kazlauskas, R., Murphy, P. T., Quinn, R. J., and Wells, R. J. (1976). *Aust. J. Chem.* **29**, 2533.
Kazlauskas, R., Lidgard, R. O., and Wells, R. J. (1977). *Tetrahedron Lett.* p. 3183.
Kikuchi, T., Mori, Y., Yokoi, T., Nakazawa, S., Kuroda, H., Masada, Y., Kitahara, K., and Umezaki, I. (1975). *Chem. Pharm. Bull.* **23**, 690.
Kinnel, R., Duggan, A. J., Eisner, T., Meinwald, J., and Miura, I. (1977). *Tetrahedron Lett.* p. 3913.

Krejcarek, G. E., White, R. H., Hager, L. P., McClure, W. O., Johnson, R. D., Rinehart, Jr., K. L., McMillan, J. A., Paul, I. C., Show, P. D., and Brusca, R. C. (1975). *Tetrahedron Lett.* p. 507.
Kurata, K., and Amiya, T. (1975). *Nippon Suisan Gakkaishi* **41**, 657.
Kurata, K., and Amiya, T. (1977). *Chem. Lett.* p. 1435.
Kurata, K., Amiya, T., and Yabe, K. (1973). *Nippon Suisan Gakkaishi* **39**, 973.
Kurata, K., Amiya, T., and Nakano, N. (1976). *Chem. Lett.* p. 821.
Kusumi, T., Shibata, Y., Ishitsuka, M., Kinoshita, T., and Kakisawa, H. (1979). *Chem. Lett.* p. 277.
Lindsey, A. S. (1965). *J. Chem. Soc.* p. 1685.
Lock, G. (1930). *Monatsh. Chem.* **55**, 167.
Lovell, F. M. (1966). *J. Am. Chem. Soc.* **88**, 4510.
Low, E. M. (1951). *J. Mar. Res.* **10**, 239.
Lowe, G. (1965). *Chem. Commun.* p. 411.
Lundgren, L., Olsson, K., and Theander, O. (1979). *Acta Chem. Scand.* **B33**, 105.
McEnroe, F. J., and Fenical, W. (1978). *Tetrahedron* **34**, 1661.
McLachlan, J., and Craigie, J. S. (1966). *J. Phycol.* **2**, 133.
McMaster, W. J., Scott, A. I., and Trippet, S. (1960). *J. Chem. Soc.* p. 4628.
Manley, S. L., and Chapman, D. J. (1978). *FEBS Lett.* **93**, 97.
Manville, J. F., and Troughton, G. E. (1973). *J. Org. Chem.* **38**, 4278.
Markham, K. R., and Porter, L. J. (1969). *Phytochemistry* **8**, 1777.
Marner, F.-J., Moore, R. E., Hirotsu, K., and Clardy, J. (1977). *J. Org. Chem.* **42**, 2815.
Martín, J. D., and Darias, J. (1978). In "Marine Natural Products, Chemical and Biological Perspectives" (P. J. Scheuer, ed.), Vol. I, pp. 125–173. Academic Press, New York.
Minale, L. (1978). In "Marine Natural Products, Chemical and Biological Perspective" (P. J. Scheuer, ed.), Vol. I, pp. 175–240. Academic Press, New York.
Minale, L., Riccio, R., and Sodano, G. (1974). *Tetrahedron Lett.* p. 3401.
Minale, L., Cimino, G., De Stefano, S., and Sodano, G. (1976). *Progr. Chem. Org. Nat. Prod.* **33**, 1.
Montaudo, G., Finochiaro, P., Trivellone, E., Bottino, F., and Maravigna, P. (1971). *Tetrahedron* **27**, 2125.
Moody, K., Thomson, R. H., Fattorusso, E., Minale, L., and Sodano, G. (1972). *J. Chem. Soc. Perkin I* p. 18.
Moore, R. E. (1977). *Accounts Chem. Res.* **10**, 40.
Moore, R. E. (1978). In "Marine Natural Products, Chemical and Biological Perspectives" (P. J. Scheuer, ed.), Vol. I, pp. 43–124. Academic Press, New York.
Mörner, C. Th. (1913). *Hoppe-Seyler's Z. Physiol. Chem.* **88**, 138.
Mynderse, J. S., and Moore, R. E. (1978). *J. Org. Chem.* **43**, 2301.
Mynderse, J. S., Moore, R. E., Kashiwagi, M., and Norton, T. R. (1977). *Science* **196**, 538.
Ochi, M., Kotsuki, H., Muraoka, K., Tokoroyama, T. (1979). *Bull. Chem. Soc. Jpn.* **52**, 629.
Ogino, C., and Taki, Y. (1957). *Tokyo Univ. Fish.* **43**, 1.
Ohta, K., and Takagi, M. (1977a). *Phytochemistry* **16**, 1062.
Ohta, K., and Takagi, M. (1977b). *Phytochemistry* **16**, 1085.
Ohtsuru, M., Teraoka, M., Tori, K., and Takeda, K. (1967). *J. Chem. Soc. B* p. 1033.
Pedersén, M. (1978). *Phytochemistry* **17**, 291.
Pedersén, M., and Dasilva, E. J. (1973). *Planta* **115**, 83.
Pedersén, M., and Fries, L. (1975). *Z. Pflanzenphysiol.* **74**, 272.
Pedersén, M., Saenger, P., and Fries, L. (1974). *Phytochemistry* **13**, 2273.
Prota, G., Ito, S., and Nardi, G. (1977). In "Marine Natural Products Chemistry" (D. J. Faulkner and W. H. Fenical, eds.), p. 45. Plenum Press, New York.
Pummerer, R., Puttfarcken, H., and Schopflocher, P. (1925). *Chem. Ber.* **58**, 1808.

Ragan, M. A., and Craigie, J. S. (1976). *Can. J. Biochem.* **54,** 66.
Saito, T., and Ando, Y. (1955). *J. Chem. Soc. Jpn.* **76,** 478.
Sakakibara, M., and Matsui, M. (1979). *Agric. Biol. Chem.* **43,** 117.
Sattler, E., and Glombitza, K.-W. (1975). *Arch. Pharm.* **308,** 813.
Sattler, E., Glombitza, K.-W., Werli, F. W., and Eckhardt, G. (1977). *Tetrahedron* **33,** 1239.
Scarpati, M. L., and Oriente, G. (1958). *Tetrahedron* **14,** 43.
Scheuer, P. J. (1973). "Chemistry of Marine Natural Products." Academic Press, New York.
Scheuer, P. J. (1977). *Israel J. Chem.* **16,** 52.
Schmitz, F. J., Campbell, D. C., Hollenbeak, K., Vanderah, D. J., Ciereszko, L. S., Steudler, P., Ekstrand, J. D., Van der Helm, D., Kaul, P., and Kulkarni, S. (1977). *In* "Marine Natural Products Chemistry" (D. J. Faulkner and W. H. Fenical, eds.), p. 293. Plenum Press, New York.
Seikel, M. K., Chow, J. H. S., and Feldman, L. (1966). *Phytochemistry* **5,** 439.
Sharma, G. M., and Vig, B. (1972). *Tetrahedron Lett.* p. 1715.
Sharma, G. M., Vig, B., and Burkholder, P. R. (1970). *In* "Food-Drugs from the Sea Proceedings 1969" (H. W. Youngken, Jr., ed.), p. 307. Marine Technology Society, Washington, D. C.
Sheikh, Y. M. (1969). PhD Dissertation, Univ. of Hawaii.
Sheikh, Y. M., and Djerassi, C. (1975). *Experientia* **31,** 265.
Shirahama, K. (1942). *J. Fac. Agric. Hokkaido Imp. Univ.* **49,** 57.
Sieburth, J. M., and Jensen, A. (1969). *J. Exp. Mar. Biol. Ecol.* **3,** 275.
Sims, J. J., Fenical, W., Wing, R. M., and Radlick, P. (1971). *J. Am. Chem. Soc.* **93,** 3774.
Sims, J. J., Rose, A. F., and Izac, R. R. (1978). *In* "Marine Natural Products, Chemical and Biological Perspectives" (P. J. Scheuer, ed.), Vol. II, pp. 297–378. Academic Press, New York.
Sleeper, H. L., and Fenical, W. (1977). *J. Am. Chem. Soc.* **99,** 2367.
Stallard, M. O., and Faulkner, D. J. (1974a). *Comp. Biochem. Physiol.* **49B,** 25.
Stallard, M. O., and Faulkner, D. J. (1974b). *Comp. Biochem. Physiol.* **49B,** 37.
Stempien, M. F., Jr., Chib, J. S., Nigrelli, R. F., and Mierzwa, R. A. (1973). *In* "Food-Drugs from the Sea Proceeding 1972" (L. R. Worthen, ed.), p. 105. Marine Technology Society, Washington, D. C.
Stoffelen, H., Glombitza, K.-W., Murawski, U., and Bielaczek, J. (1972). *Planta Med.* **22,** 396.
Stonard, R. J., and Andersen, R. J. (1980). *3rd Intern. Symp. Mar. Nat. Prod., 1980,* P 5.
Suzuki, M., and Kurosawa, E. (1976). *Tetrahedron Lett.* p. 4817.
Suzuki, M., and Kurosawa, E. (1978). *Tetrahedron Lett.* p. 2503.
Suzuki, M., Hayakawa, Y., and Irie, T. (1969). *Bull. Chem. Soc. Jpn.* **42,** 3342.
Takahashi, T. (1931). *Tokyo Kogyo Shikensho Hokoku* **26,** 1.
Thomson, R. H. (1971). "Naturally Occurring Quinones," 2nd ed. Academic Press, New York.
Tocher, R. D., and Craigie, J. S. (1966). *Can. J. Bot.* **44,** 605.
Tsuruta, H., and Mukai, T. (1968). *Bull. Chem. Soc. Jpn.* **41,** 2489.
Waraszkiewicz, S. M., and Erickson, K. L. (1974). *Tetrahedron Lett.* p. 2003.
Weber, K., and Ernst, W. (1978). *Naturwissenschaften* **65,** 262.
Weinstein, B., Rold, T. L., Harrell, Jr., C. E., Burns, M. W., III, and Waaland, J. R. (1975). *Phytochemistry* **14,** 2667.
Yamamura, S., and Hirata, Y. (1963). *Tetrahedron* **19,** 1485.

Chapter 4

Marine Chemical Ecology: The Roles of Chemical Communication and Chemical Pollution

MICHEL BARBIER

MARINE NATURAL PRODUCTS

ISBN 0-12-624004-3

Recent rapid advances in the chemistry of natural products are due to a large extent to technical progress in structure determination. Structures can now be established with less than one milligram of material—and quite rapidly if suitable crystals can be obtained. Despite these developments, marine chemistry had long been a neglected segment of natural-products research. Its current popularity tends to let us forget the fact that for years contributions were made by a few organic chemists who tried to relate observed biological phenomena to the molecules they isolated. Some of these chemists performed true pioneering work, often under difficult external conditions. Volume II of this series drew particular attention to the role of Professor Edgar Lederer in the early development of this field. As one of his collaborators for about 30 years, I wish to express my admiration and gratitude. Under his guidance, interest in marine chemistry and biochemistry was fostered in our laboratory. From our early study of calliactine, a red pigment from the sea anemone *Calliactis parasitica,* first described by Abeloos and Teissier (1926) and isolated crystalline by Lederer *et al.* (1940), our interest in products of marine origin has never diminished. Our research later led us to explore the possible ecological significance of the identified substances, and thus formed the nucleus of chemical ecology.

I. INTRODUCTION

Interactions of all kinds, including chemical interactions, are a fundamental characteristic of nature. In our search for a definition of natural phenomena we can do no better than finding the molecular basis of these complex relationships. We discover the existence of interspecific cybernetics, the logic of which contrasts with the apparent randomness that we observe. Consequently, the notion of adaptive advantage becomes important when we look for the significance of observed relationships.

For this chapter, several examples have been chosen which help to classify chemical interactions and which demonstrate some peculiarities due to the marine environment. Thus, the role of such molecules as sterols and growth factors dissolved in seawater and the concepts of antibiosis and symbiosis will be emphasized, along with some examples of ecological toxins. The mercury cycle and hydrocarbon spillage will be considered briefly. This consideration introduces the important capacity of some marine organisms to concentrate matter and leads to a discussion of the consequences of this property.

For a long time to come, the marine environment will remain a source of new molecular structures and of newly discovered biological activities, thus ensuring motivation for research. Chemical ecology is the key to the search for these molecules. It also provides the opportunity for a reconsideration of their significance in adaptive processes which ultimately govern the path of evolution. In fact, chemical ecology may be considered as an ambitious and unique enterprise, which makes the work of the chemist meaningful.

II. DEFINITIONS AND GENERAL CONSIDERATIONS

One exciting recent discovery is the demonstration of the probability of a prebiotic era during which chemical exchanges took place on the surface of mineral or organic polymers, thus pointing to a prehistoric model of a membrane. This discovery certainly introduces a somewhat mechanistic consideration to the origin of life, in agreement with Herodotus: "Given enough time, everything possible will happen," via a process that we call evolution. The permanence of organic molecules, particulate or dissolved, has been and remains the basis of life. From this basis there follow the complex functions of adsorption and excretion—first on mineral supports, then on some giant molecules like Oparine's aggregates and, in turn, primitive cells—which lead to the elementary principles of chemical communication. Oparine, as did Claude Bernard, recognizes the difficulty in defining life itself correctly, and insists on a common feature of all organisms, namely, the need to communicate.

Since the proposals by Sondheimer and Simeone (1970) and the classification of Whittaker and Feeny (1971), which introduced the term chemical ecology[1] and a series of new definitions, communication by chemical substances has become an accepted concept. Yet many chemists seem to avoid the proposed terminology.

Ecology is in fact the science of the relationship between organisms and their environment. Former classifications distinguished between ecology *sensu stricto*, demecology [population dynamics], and synecology [relationships between communities]. Chemical ecology, rather, should be derived from biocenology, which deals with the inter- and intraspecific interactions in nature.

The study of chemical ecology may be viewed as an attempt to transform the chemistry of natural substances into more systematic research. Search for proper terminology, together with the creation of new words,

---

[1] Books on chemical ecology are available: Sondheimer and Simeone (1970), Barbier (1976, 1978, 1979), and Harborne (1977).

is inherent in an incipient science. Classification of interactions in nature makes structural elucidations meaningful, and provides for the chemist new leads in his quest for original investigations.

Perhaps the natural phenomena most difficult to understand are those that are part of cycles and participate in multiple equilibria and thus form the basis of an ecosystem. The continuous exchanges of molecules between the inorganic and organic spheres, coupled with the continuous interactions among species, are somewhat reminiscent of the pre-Socratic philosophies in which cycles of opposite processes play a prominent part. These thoughts were translated into scientific terms by Lavoisier in a communication to the Paris Academy of Sciences as early as 1792. Lavoisier wrote about the "circulation" of molecules among the three kingdoms, using the words "animalization" and "vegetation" as opposed to "combustion" and "putrefaction." These were highly advanced ideas on ecology and biochemistry. Lavoisier's communication was so important that it formed the basis of Pasteur's work.

The problem for many organisms is not to find a way to live better, but to find a way to survive under difficult conditions, a search that results in biochemical evolution. Organic molecules act in the evolutionary scheme as tools, with the earth as support, and air or water as media for transportation. Somehow, interactions in liquid media have not been subjected to the same intensive research as have terrestrial phenomena, and chemical ecology of the marine environment remains the stepchild of chemical ecological research. As will be seen in the following, marine chemistry has been extensively developed in the last few years, but not always in the direction of chemical ecology, which is still rather neglected. A persuasive example of this neglect is our lack of knowledge of marine pheromones.

A. Systematic Classification of Interactions

The natural multiplicity of chemical interactions, which reaches a maximum in the liquid phase, may be analyzed in simple terms, as shown in Figs. 1–3. Inorganic and organic molecules circulate in the three kingdoms, mineral, plant, and animal (Fig. 1). The producer may be of the same species as the receiver (intraspecific) or of a different species (interspecific) (Fig. 2). Whether the molecule produced gives an adaptive advantage to the producer or to the receiver is an important consideration (Fig. 3) for classification. The Whittaker and Feeny proposals (1971) based on this scheme are presented in Table I. The interspecific (*allelochemic*) effects are divided into the *allomones*, which give an adaptive advantage to the producing organisms, and *kairomones*, which give the

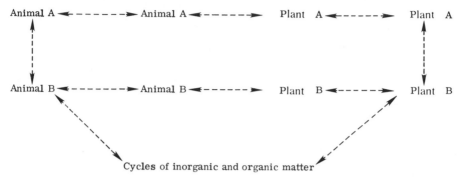

Fig. 1. Simplified scheme of possible interactions in nature.

Interspecific   A — — — — — ➤ B   Allelochemic effects

Intraspecific   A — — — — — ➤ B   Pheromones, for example

Fig. 2. The two cases of plant or animal interactions in nature.

advantage to the receiver (see also Fig. 3). The notion of adaptive advantage is important as it emphasizes the significance of organic molecules in their natural environment. Among the intraspecific effects are found a wide group of pheromones, the definition of which has to be discussed. The term "pheromone" was coined by Karlson and Lüscher in 1959 from the Greek *pherein* [to transport] and *horman* [to excite], a combination which should have led to the spelling "pherormone," which is sometimes encountered in the literature. According to these authors, pheromones are substances secreted to the outside of an individual and received by an organism of the same species in which it triggers a specific reaction, a definite behavior or a developmental process (biochemical modification). It is necessary to emphasize the fact that the word, first applied to insects, was not limited to them by definition. Astonishingly enough, lately the term "gamone" has been preferred for the characterization of compounds responsible for the attraction between gametes of fungi or algae (e.g., spermatozoids attracted to the eggs) (Reschke, 1969; Raper, 1952;

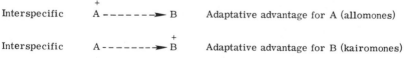

Fig. 3. The two cases of plant or animal interactions in nature.

TABLE 1

Classification of Different Types of Interactions[a]

I. Allelochemic effects (interspecific)
  A. Allomones: Adaptative advantage to the producer
    1. Repellents
    2. Escape substances
    3. Suppressants and antibiotics
    4. Venoms
    5. Inductants (leading to galls, nodules, etc.)
    6. Counteractants (neutralizing agents, antibodies)
    7. Attractants (to a predator, pollination attractants, etc.)
  B. Kairomones: Adaptative advantage to the receiver
    1. Food location signals
    2. Inductants stimulating a particular development (spine-factor in rotifers)
    3. Signals warning of a danger or toxicity, predator scents
    4. Stimulants such as products inducing growth in the receiver
  C. Depressants; wastes: No adaptative advantage
II. Intraspecific chemical effects
  A. Autotoxins
  B. Autoinhibitors, fixing a limit to populations (spermatocides, staling substances of fungi)
  C. Pheromones with roles in reproduction, social regulation, control, recognition, alarm, territory marking, trail marking, food marking and location

[a] From Whittaker and Feeny (1971). Copyright © 1971 by the American Association for the Advancement of Science.

Jaenicke and Müller, 1973). For the same intraspecific molecules the American literature has chosen "hormone."

Transmission of information by chemicals is a sophisticated part of cybernetics. Law and Regnier (1971) proposed the name "semiochemical" for each molecule possessing such a property. In fact, if one looks at the classification of effects, one is faced with a determinant aspect of nature, a multidimensional network of relations between cause and effect that is based on complex natural phenomena. Aubert et al. (1972) and Aubert (1976) proposed a similar concept and introduced the term "telemediators" for compounds acting in the sea. He suggested the following definition: "Telemediators are substances synthesized by marine animal or plant species which, when released into the environment, act remotely upon the behavior or the biological functions of the same species or other species, excluding chemical actions due to the environment itself."

Kirschenblatt's earlier terminology (1957) fits the aforementioned considerations (Figs. 2 and 3). His term "telergone" indicates a chemical action at a distance, with prefixes "homo" or "hetero" to distinguish between intra- or interspecific effects. In spite of a defense

(Kirschenblatt, 1962) of his elaborate terminology versus the term pheromone, his nomenclature has not been accepted.

Whittaker and Feeny's classification (Table 1) clearly shows the great importance of defense mechanisms, which are directly or indirectly involved in allomones and kairomones and which translate the vital necessity to become adapted into the terms of chemical phenomena. Pollutants, which constitute a particular case of depressants, are set apart as they lead to no adaptive advantage. Unfortunately (and this should perhaps be reconsidered), nothing in this scheme seems to be related to the intracellular machinery and to molecules responsible for cell life and cell division, which are often message transmitters and carry information. Florkin (1966) contributed an essential idea to this question: "it appears clearly that in the network of the biochemical continuum, is operating a circulation of specific molecules or macromolecules, endowed with a certain quantity of information."

As will be seen later, direct field interaction between marine chemists and naturalists or biologists who study marine phenomena is an absolute necessity. As has been noted by Ciereszko (1976), there will always be a serious need (and pleasure) for the chemist to go back to the beach in search of direct observation.

Chemical ecology could mark a turning point in human evolution; but its predictive value is still very limited (Eisner, 1972), and its impact on human society remains negligible in spite of an increasing interest. It is, however, still a great hope that marine chemical ecology will in the future lead to unsuspected discoveries in a field of research that appears unlimited.

B. A Simplified Scheme of Ocean Water Layers

A crucial part of the marine environment is the surface microlayer where numerous phenomena, a high concentration of unicellular organisms, and a rich accumulation of dissolved organic matter are encountered. MacIntyre (1975) states that "the chemistry of three-quarters of the planet is found at the sea-air interface, which has been for a long time neglected and is now the object of a growing attention." All properties of this film are different from the rest of the oceans; it is here that many mechanisms (reviews in Oldham et al., 1978; Barbier and Saliot, 1976) are determined and where algal and bacterial activity is found. This "skin of the sea" should be a virtually unlimited source of investigation for the future. If the inventory of the substances dissolved in seawater at that level and their photochemistry have just begun to be studied, nearly everything remains to be done that deals with the genesis of biologically interesting products by the varied constituents of this mixture of micro-flora and microfauna.

In other parts of the sea, the concentration of dissolved organic mole-cules is poorer. It is related to the extent of the euphotic zone, i.e., where the light penetrates, or to the proximity of the sea bottom, which is a region of active exchange of matter. However, as will be seen later, 100 μg/liter of dissolved organic matter still represents a huge number of molecules, which is significant for unicellular organisms and for the con-cept of seawater. Fossils and deep-sea waters, where renewal is very slow, are of great interest as they provide a comparative basis of composi-tion for a period prior to industrial pollution. The proximity of continents, mainly around estuaries, also leads to different composition due to the input of terrestrial material. This situation has been well analyzed by Tusseau *et al.* (1978), who studied dissolved sterols of the Atlantic along the northern Brazilian coast and around the Amazon delta. The stigma-sterol concentration increases regularly as one approaches the delta. In European waters, cholesterol concentration increases near the harbors, as one would expect.

Life in the oceans is thus localized at different water levels. First there is the surface microlayer, then the rest of the euphotic zone. Other modifications are generally introduced by temperature changes, warm currents, or particular ecosystems that permit good development of many species which are often interrelated as, for example, on coral reefs.

Benthic and pelagic animals or plants represent only a small part of the living organisms but remain, because of accessibility and size, preferred targets for many scientists interested in marine chemistry. The most significant organisms, the microflora and microfauna, are awaiting isola-tion and culture.

Among a series of questions which can be asked about the marine environment are some major ones, such as the following (Fontaine, 1976): What is the historical fate of a given dissolved organic molecule in the sea? Do these dissolved substances have any direct or indirect biological activity? What is the origin of the dissolved organic matter? This is already a full program for future comprehensive chemical research of the marine environment.

III. ANTIBIOSIS IN THE MARINE ENVIRONMENT

A. Ecology of Antibiotic (versus Antimicrobial) Substances in the Marine Environment

The term antibiosis is taken from Vuillemin and was introduced as an antonym to symbiosis as early as 1887 by Pasteur and Joubert who reported the antagonistic properties of saprophytic bacteria toward the growth of *Bacillus anthracis* in culture. However, the word antibiotic was only coined in 1940 by Waksman and was reserved for substances produced by microorganisms that, at low concentrations, inhibit the development of microorganisms of other species. By the Whittaker and Feeny classification, with the adaptive advantage in favor of the producer, antibiotics are typical examples of allelochemicals. If one ignores the significance of the active compounds for the microorganism itself, an antibiotic is an allomone but could perhaps be, at least in some cases, an autoregulator or a waste product. In fact, the definition has been considerably broadened as it is now being applied to many kinds of growth-inhibiting products, even if they are not produced by microorganisms and even when they are obtained by synthesis without having a natural counterpart. Why should this be of concern, as it is only a question of definition? Unfortunately, antibiotic activities are often extrapolated from mere bacteriostatic effects. Yet, when a metabolite has not evolved in order to maintain the ecological niche of an organism (the adaptive advantage), its interaction on the growth of another organism is incidental. By contrast, when one looks for new antibiotics, a search for interactions and space competition among species may be a real path to discovery.

In the marine literature the term antibiotic has sometimes been used equally casually. Most compounds have been tested for antimicrobial activity *in vitro* only. This has led some authors to a careful appraisal of the situation (see, for example, Grant and Mackie's (1977), article, "Drugs from the Sea, Fact or Fantasy"). For instance, a broad spectrum of phenolic substances, with or without halogen, has been isolated from various marine organisms. These compounds have long been known to pharmacologists as good external bacteriostatic substances. On the other hand, marine chemical ecologists are interested in these molecules (more, perhaps, than biologists) in order to look at the possible summation of the activities in the environment and to reanalyze the observed interactions in terms of adaptation. Reviews that deal with such "antibiotics" from marine animals or plants are numerous (Baslow, 1969; Li *et al.*, 1974; Premuzic, 1971; Sims *et al.*, 1975; Webber and Ruggieri, 1974; Rinehart *et al.*, 1976, Faulkner, 1979). The list of new chemical structures is of great interest, but most of these substances will find no practical application in human therapy.

B. From Iodoform to the Arabinosyl Nucleosides: The Diversity

Some simple molecules, such as iodoform and bromoform produced by algae (McConnell and Fenical, 1977; Codomier *et al.*, 1977), have strong bacteriostatic activity; iodoform itself has been used extensively for treatment of wounds. Many metabolites of the red algal family Bonnemaisoniaceae supposedly possess similar activity, among them the haloketones, the haloacrylic acids, chloro- and bromoalcohols, etc. The natural mixture is supposed to render the algae more resistant and quite unpalatable to phytophagous invertebrates. In the vicinity of stands of these algae the accumulation of these molecules dissolved in seawater should play a role in the equilibria of species and should participate in the conditioning of the environment. This is also the case for acrylic acid, which is produced by several species of algae (Grant and Mackie, 1977; Glombitza, 1970) and which has marked antimicrobial activity. The vicinity of macroalgal growth always has a higher concentration of dissolved polyphenolic substances, most of them of the phloroglucinol type (Glombitza et al., 1973; Glombitza, 1977). These substances may interfere with the growth of unicellular planktonic organisms and moderate their development. Ecological model studies are needed in all these relationships in order to establish parameters of the phenomena and of their consequences.

Systematic searches for "antibiotic" properties have sometimes led to curiosities, for instance, the unexpected isolation of 3,5-dinitroguaiacol or dichloroacetamide from algae (Ohta and Tagagi, 1977).

Among molecules with a more sophisticated structure are the fimbrolides (Kazlauskas et al., 1977), the beckerelides (1) (Ohta, 1977), or laurinterol (2) and cycloeudesmol (3), which have substantial in vitro antibiotic activity, but in vitro only. Laurinterol (Sims et al., 1975) from the alga Laurencia pacifica has the same activity as streptomycin on Staphylococcus aureus and Mycobacterium smegmatis (1 to 5 μg/mL for complete growth inhibition after 48 h), but this compound does not seem to be clinically applicable. Cycloeudesmol has the property of inhibiting some Gram-negative microorganisms, at least in the test tube. Such activities, however, are unexpected of relatively simple terpenes.

Glycosidic substances such as the asterosaponins from starfish, e.g., 4 (Schmitz, 1978), or the holothurins from holothurians (5) (Kitagawa et al., 1979) exhibit, in addition to their repellent effects on mollusks, antibacterial and antiviral properties which may be related to a general toxicity, thus prohibiting until now any in vivo application. The violent escape response of mollusks (Pecten) to the asterosaponins indicates good adaptation to a predator (kairomonal reaction).

Isolation of cephalosporin C (6) (Abraham, 1967) in cultures of a fungus collected from a sewage outfall in the sea is not really a case of marine antibiosis, but is interesting as it indicates possible interactions of terrestrial microorganisms with the coastal environment.

Thelepin (7) (Higa and Scheuer, 1975) has been found in the polychaete

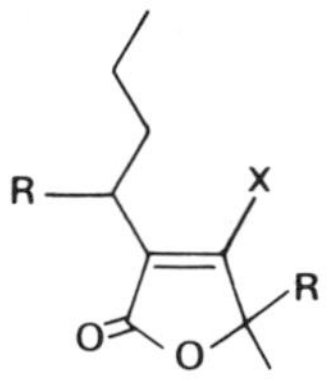

R
OH
OH

X
Br
Cl

R_1 = SO_3H

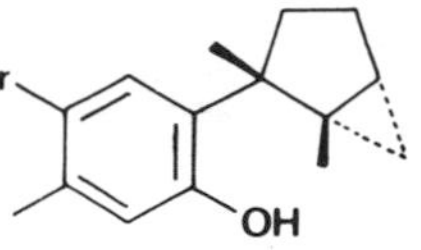

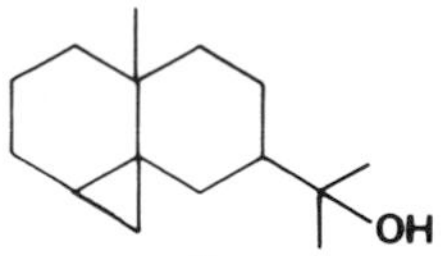

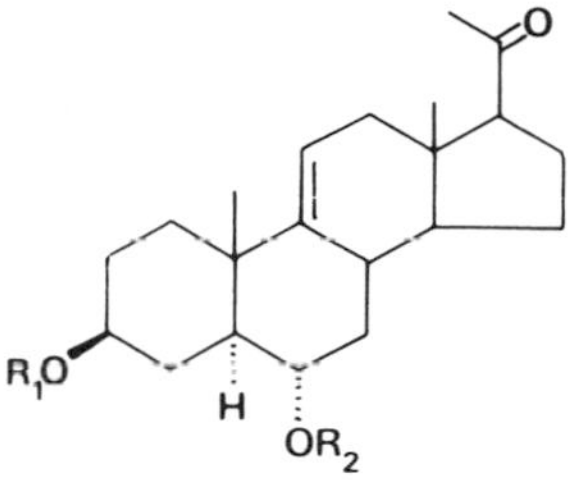

R_2 = 2 moles fucose
       + 2 moles quinovose

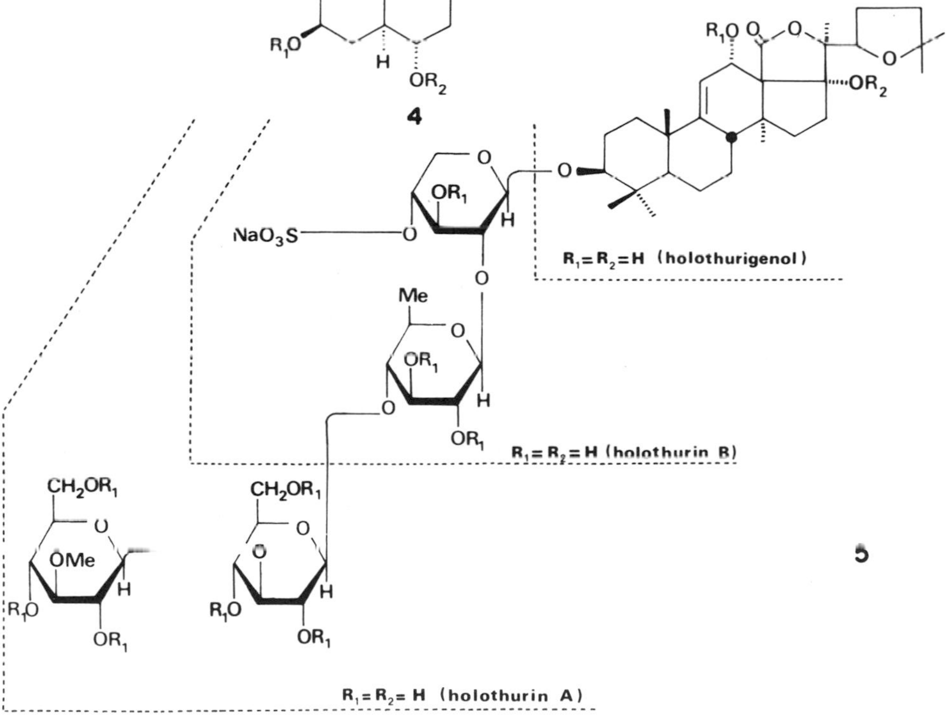

4

$R_1 = R_2 = H$ (holothurigenol)

NaO_3S

$R_1 = R_2 = H$ (holothurin B)

CH_2OR_1

OMe

R_1O

OR_1

CH_2OR_1

R_1O

OR_1

5

$R_1 = R_2 = H$ (holothurin A)

worm *Thelepus setosus* and is a powerful antifungal agent, of the same potency as griseofulvin of similar structure.

Curiously enough, the most promising growth-inhibiting molecules ever isolated from the marine world were also among the first products found, namely, the unusual nucleosides from the Caribbean sponge *Cryptotethia crypta*. Spongothymidine (**8**) and spongouridine (**9**), which are 1-β-arabinosyl derivatives of the corresponding bases (Bergmann and Feeny,

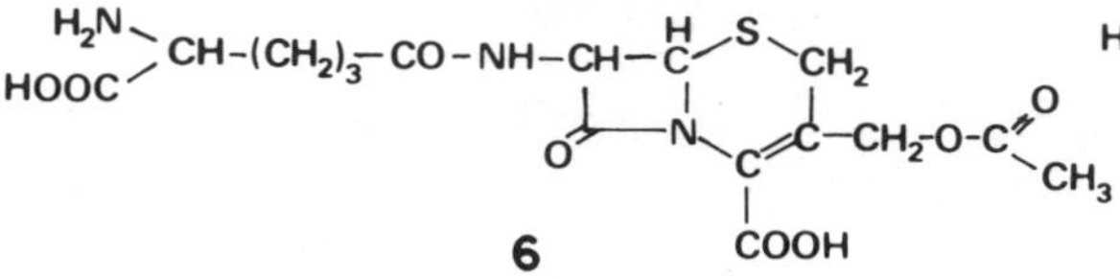

6

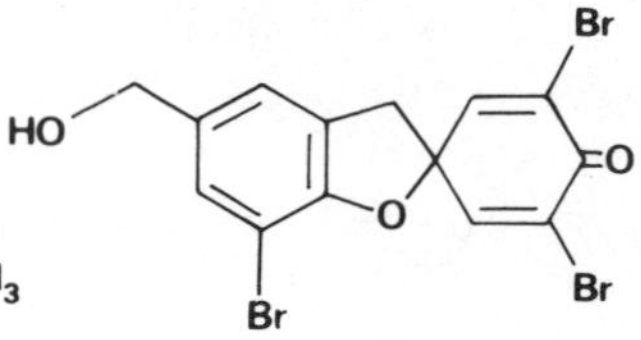

7

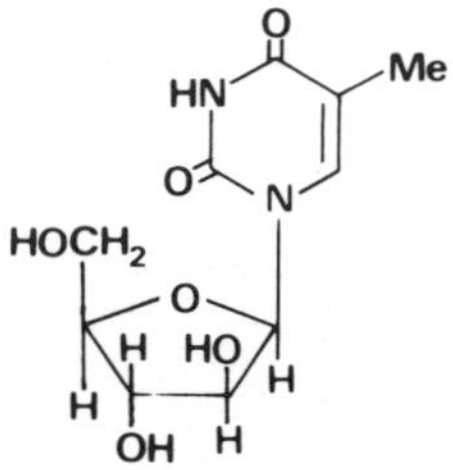

8

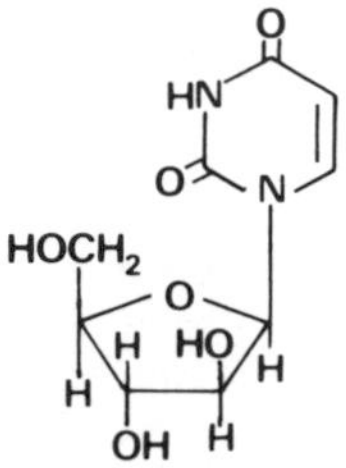

9

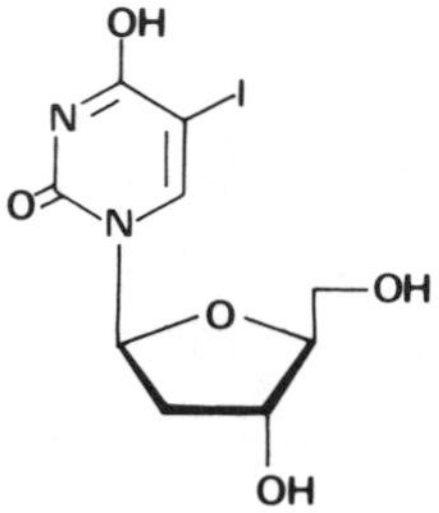

10

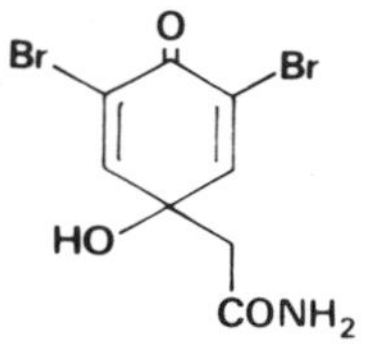

11

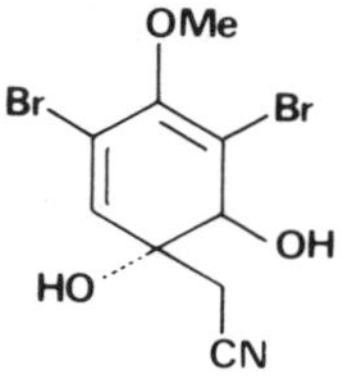

12

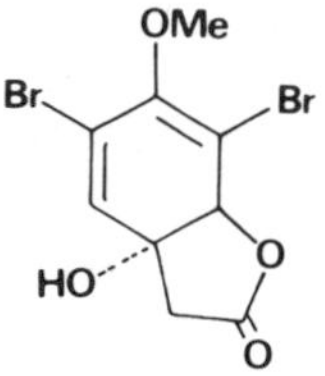

13

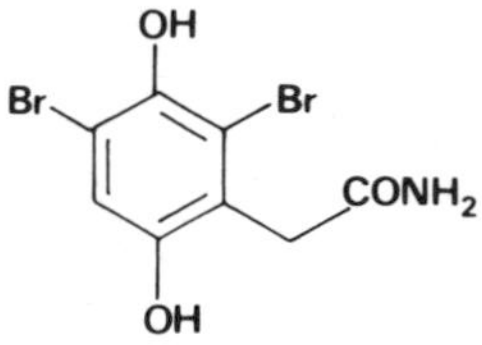

14

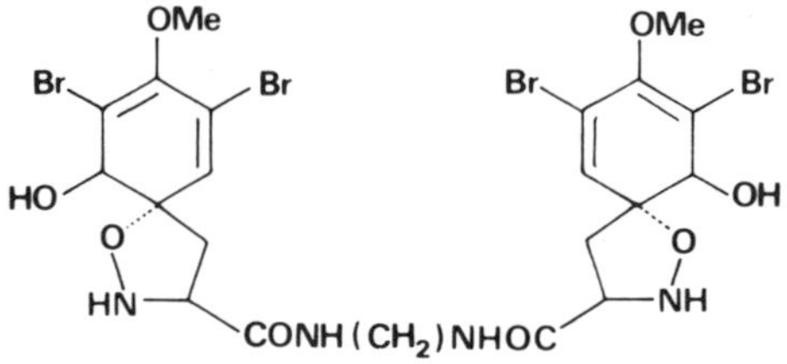

15

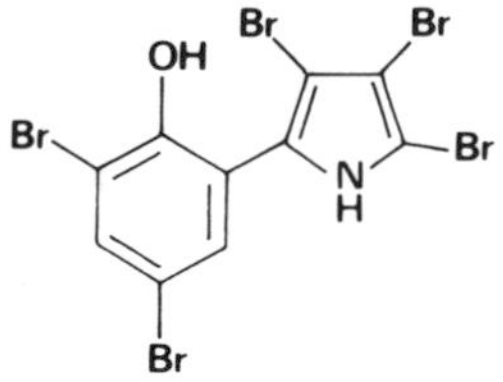

16

1950, 1951), have led to a powerful synthetic product 1-β-2'-deoxyribofuranosyl-5-iodouracil (**10**). Now a commercial product, compound **10** has been successful (Bloch, 1975) against viral infections and, more particularly, against the dangerous *herpes* virus. Clinical studies are now being conducted on adenine arabinoside and other derivatives, which are important tools in cancer research. Adenine arabinoside inhibits the biosynthesis of DNA through its ready incorporation at the terminal of DNA chains during elongation. This produces fragments of no genetic significance. Adenine arabinoside has no immunosuppressive effect and does not lead to resistance, but, when applied together with cytosine arabinoside (Privat de Garilhe and De Rudder, 1970; Privat de Garilhe, 1977), performs more effectively, which may lead to possible new therapeutic strategies. These substances seem to be the only compounds resulting from marine research on antibiotic or antiviral activities that have medicinal application. Numerous developments have taken place in the synthetic field for the preparation of other suitable agents for cancer research.

C. Dual Aspects of the Search for Antimicrobial Activity in the Marine Environment

Investigations of marine organisms in search of antimicrobial activity began only 7 years ago. As about 500,000 species (in 30 phyla) must be explored, there is still an enormous task ahead. Moreover, many bacteria, fungi, and unicellular algae, for example, are not available to the chemist unless the organisms are grown in culture. This will be necessary for future research. Since a majority of strains have not yet been isolated, this task must begin with oceanographers and microbiologists before chemists can do significant research. The ocean surface microlayer, the euphotic zone, and the upper part of the sediments are particularly rich in microorganisms of all kinds which are awaiting isolation.

The competing-systems or caged-cultures techniques (Jensen *et al.,* 1972) may give rise to antagonists through adaptation to an unusual condition. In these systems, cultures of two organisms are maintained on both sides of a diaphragm, permitting dialysis of molecules and hence, presumably, competition. These model experiments, which permit interfering conditions, offer open-ended programming and could be the source of future serendipitous discoveries. Artificial allelopathic phenomena can thus be produced in the laboratory.

Interacting metabolisms are probably involved between many marine invertebrates and symbiotic algal colonies, but this idea also remains hypothetical in most cases and must be proved. Among pertinent examples, let us consider the origin of halogenated derivatives in invertebrates or of sterols with a permethylated side chain. (As discussed in the following, invertebrates are generally unable to biosynthesize such sterols.) The dibromo aromatic compounds, **11–14,** which were isolated from sponges (Faulkner, 1977; Moody *et al.*, 1972), or aerothionin (**15**) from the sponge *Verongia,* which show *in vitro* bacteriostatic activity, might have originated from dibromotyrosine, a common algal amino acid, but this assumption also awaits experimental verification.

These examples only serve to underline the lack of experimental work at the interface of organic chemistry and ecology and the necessity of progressing in the direction of marine chemical ecology in order to gain a better understanding of natural phenomena.

The search for antibiosis, or for any recognizable biological activity, will remain a relatively easy way to do research and will logically lead eventually to molecules that possess the expected biological activity. The unknown activity is perhaps the most important key to the future, and the condition *sine qua non* for future progress in science will be, as ever, direct observation. Thus, even if systematic screening of species for antibiosis is still an urgent task, the slow, difficult, and uncertain process which consists of biologist–chemist interaction, of an understanding of the phenomena, and of designing a bioassay will always be the Cartesian path toward original results. Thinking about nature in terms of interacting systems led to chemical ecology. Reconsideration of the Whittaker and Feeny classification, or of any other attempt at rationalization, may possibly lead to a different system.

However, a rapid look at reviews in the field of marine natural products reveals a great variety of molecules, and this of course is still a source of motivation for the chemist—structure for structure's sake and nothing else. As expressed by Scheuer (1973), this combines brilliance with nature and is particularly true in marine chemistry, which is an open catalog of unbelievable molecular architecture, an alluring subject for the organic chemist in quest of originality. These then are the dual aspects of research in the chemistry of natural products, the search for antibiotic versus antimicrobial compounds in the marine environment.

D. Halogenated Marine Molecules and Their Ecological Significance

It is interesting that some bacteria follow similar mechanisms of halogen fixation as are encountered in algae; e.g., the pentabromophenol (16), which has antibiotic activity *in vitro* (Faulkner, 1977; Lovell, 1966), obtained from *Pseudomonas bromoutilis* and from a *Chromobacter*. To the question of whether there is a unique marine biochemistry, the answer should perhaps be no, as uniqueness appears everywhere in the world. However, some peculiarities can be recognized, and this is the case with halogen fixation. Very little is known about the mechanisms, the process of concentrating halogen from seawater, the transport and introduction of halogens into organic molecules, or the eventual selection by enzymatic systems among chlorine, bromine, and iodine. Some simple, yet sophisticated molecules found in algae, e.g., bromochloroiodomethane, would seem incredible unless they are thought of in terms of adaptation. Also fascinating is the discrimination exhibited by molecules toward halogen fixation. Many marine organisms contain halogenated terpenes, but—at least so far—no halogenated sterols. Once again, the choice of whether these molecules are biosynthesized by the invertebrates from which they were isolated, result from symbiosis, or are transformation products transmitted by the food chain is in most cases a matter of speculation.

Early in evolution, production of halogenated products by phytoplankton conditioned marine life and determined equilibria, thus introducing a kind of order into complexity. The rate of photosynthesis in the oceans is now about 10 times that of terrestrial plants. Methyl chloride from phytoplanktons, which has accumulated for centuries in the upper atmosphere, has governed (Lovelock, 1975) the conditions for UV light filtration and fixed the ozone–oxygen ratio. The ecological impact of these molecules is always important. Seawater contains on the average about 0.1 μl/liter methyl chloride (by volume), and the atmosphere immediately above sea level contains 0.2 μl/liter of total halogenated products, which participate in conditioning the air–sea interfacial zone. If in these halocarbon mixtures methyl iodide, methyl chloride, and methyl bromide are the main halogenated compounds, a marked difference exists with the air–land interface, where the atmosphere contains mainly the Freons which are man-made.

Antibiosis in the oceans is, as ever, an ecological consequence of the parameters imposed by life in a liquid medium and by the numerous resulting interactions. In such a scheme halogenated molecules produced by macro- and microalgae and by bacteria have formed the boundary conditions for the development of many other microorganisms and of invertebrates, thereby giving the whole ecosystem an appearance of a multidimensional network.

IV. MARINE STEROLS AND THEIR KAIROMONAL POSITION

A. Sterol Side-Chain Methylation and Demethylation Patterns

Among the first demonstrations of C-24 methylation leading to C_{29} sterols was the research performed by Villanueva et al. (1964) on the brown alga *Laminaria saccharina* using *S*-adenosylmethionine as methyl donor. Several laboratories have contributed to the establishment of the mechanisms involved, generally starting from cycloartenol (17) (Lederer, 1969). The reactions may occur on partially demethylated cycloartenol or on lanosterol intermediates, which lead to a number of interrelated possibilities. If C-24 methylation does not occur, opening of the cyclopropane ring accompanied by demethylations at C-4 and C-14 (not necessarily in this order) gives rise to plant cholesterol. These reactions and their mechanisms have recently been reviewed by Goad (1978) and by Nes and

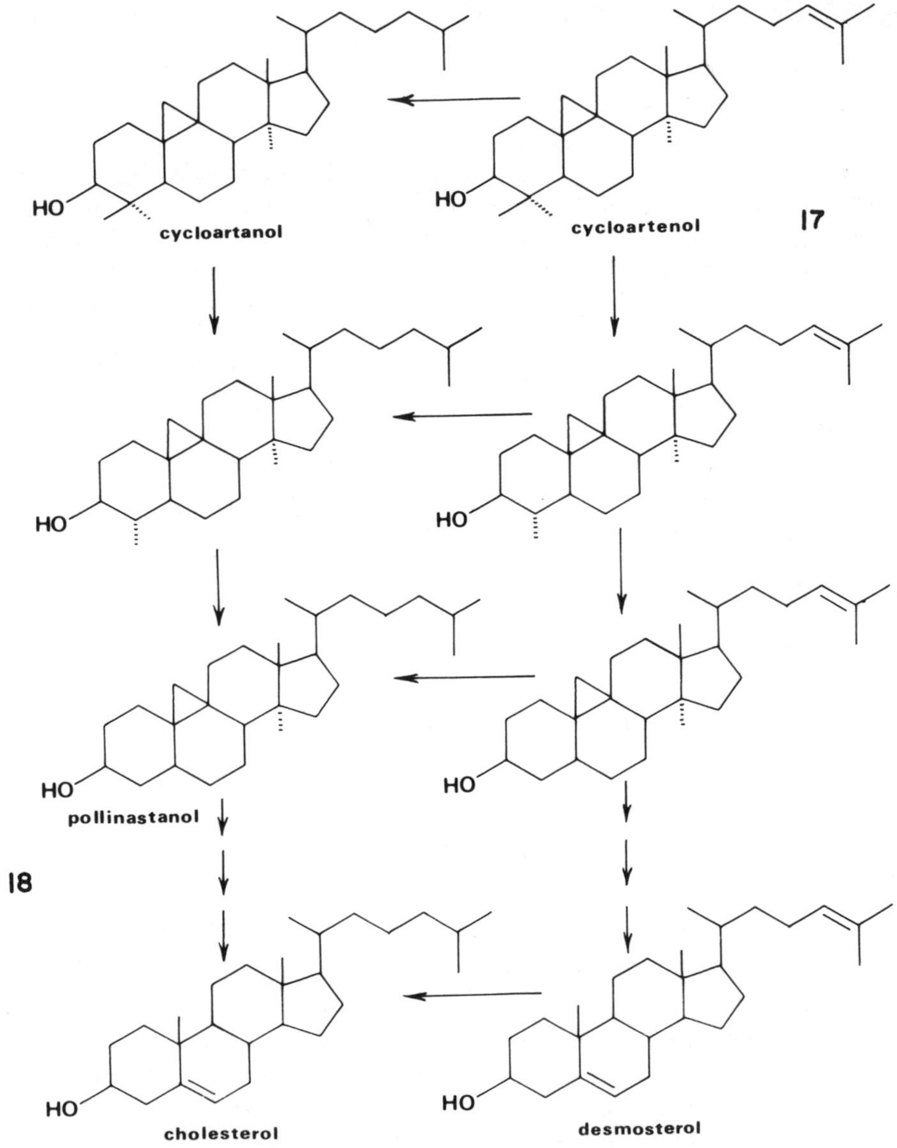

Fig. 4. The pollinastanol pathway to plant cholesterol.

McKean (1977). The multiplicity of sterol biosynthetic pathways and their corresponding bioecological implications have been considered by Barbier (1976, 1978, 1979).

If total demethylations of cycloartenol (17) at C-4 intervenes before the opening of the cyclopropane ring, the pollinastanol (18) pathway (Fig. 4) to cholesterol is initiated. Pollinastanol, first isolated from pollen and ferns (Hügel *et al.*, 1964; Devys and Barbier, 1967; Devys *et al.*, 1969a, b) has subsequently been found in different plant sources (Barbier, 1966), accompanied by some C-24 alkylated intermediates, and together with cycloartenol or cycloartanol. Since the existence of the cycloartenols has also been proved in algae (see, e.g., Ferezou *et al.*, 1974), the pollinastanol pathway to cholesterol and other C-24 methylated products should also be operative. The pollinastanol family of sterols has been found in the phytoflagellate *Astasia longa* (Rohmer and Brandt, 1973) and in the alga *Chlorella emersonii* (Doyle *et al.*, 1972).

Conversely, demethylation of plant sterols at C-24 leads to cholesterol in arthropods, as shown in Fig. 5; this is an important source of this sterol in terrestrial and marine environments (Allais *et al.*, 1973; Collignon-Thiennot *et al.*, 1973; Saliot and Barbier, 1973b; Awata *et al.*, 1975; Chen *et al.*, 1975; Allais and Barbier, 1977). A discussion of the biological significance of methylation and demethylation patterns in marine algae and invertebrates follows.

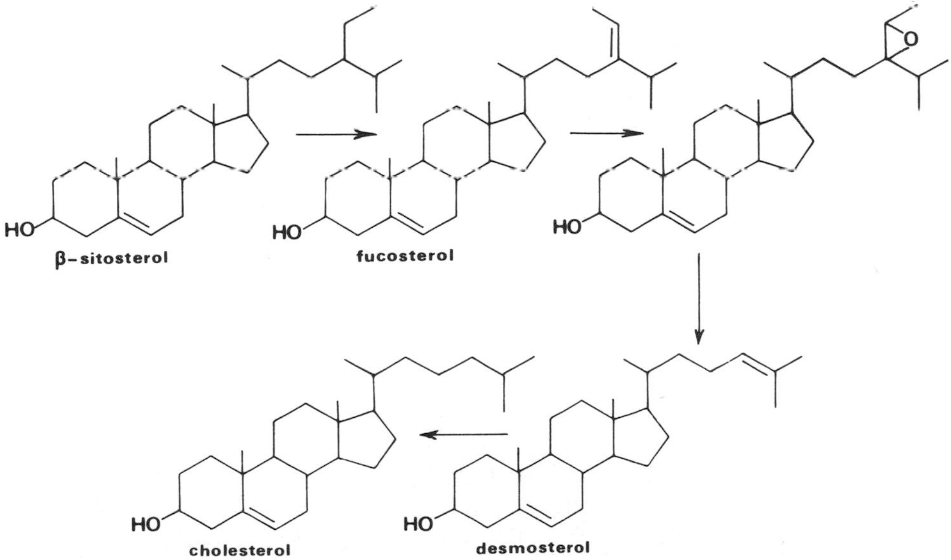

β-sitosterol   fucosterol

cholesterol   desmosterol

Fig. 5.   An arthropod source for cholesterol: the dealkylation process of plant sterols at C-24.

B. Bioecological Significance of Sterols Dissolved in Seawater

Organic molecules in the ocean, whether in solution or as particulate matter, are of interest in relation to our understanding of natural cycles, of interspecific relationships, and of geochemical evolution. All marine organisms continually exchange these molecules. These phenomena are of course vital to the smallest organisms. Planktonic unicellular species, animal or plant, are dependent on dissolved or suspended matter that is furnished by the medium. From this viewpoint the oceans represent a giant culture. Capture and metabolism of sterols, fatty acids, or hydrocarbons (Saliot and Barbier, 1973a; Février et al., 1976) have been established even for some invertebrates (e.g., sea anemone and oyster).

The composition of dissolved sterols in seawater has been subjected to several recent investigations (Gagosian, 1975; Kanazawa and Teshima, 1971; Saliot and Barbier, 1973a, b; Tusseau et al., 1978). Water samples from different locations or depths were filtered (0.45-μm filters) and extracted with a solvent (the method of the French workers); the lipid fraction was saponified and the sterols isolated on silica plates and further analyzed by gas chromatography and mass spectrometry. The methodology is important, first, because it corresponds to an agreed-upon arbitrary definition of the "dissolved" state. Second, because Δ^5-sterols are generally analyzed, with exclusion of stanols and of polyhydroxy compounds. This is somewhat prejudicial, as in the case of coprostanol, which can be used as a good indicator of pollution (Kanazawa and Teshima, 1978). Furthermore, the resulting data represent minimum concentrations since extraction yields or losses are not known.

The total amount of dissolved sterols in the Atlantic Ocean varies from 2 to 14 μg/liter (Saliot and Barbier, 1973a, b) and in the Pacific, off the Guyanas and northeastern Brazilian coasts, from 0.16 to 1.24 μg/liter, depending on the location. Contributions from the continents are clearly manifested in water samples collected near harbors where the total amount of dissolved sterols reaches 150 μg/liter, as at La Rochelle on the French Atlantic coast (Boutry and Barbier, 1974).

Cholesterol is a key metabolite for almost all species, with the possible exception of bacteria. It is found as a membrane constituent, in fats where it is esterified with fatty acids, or as an intermediate in steroid biosynthesis (e.g., the ecdysteroids of crustaceans). Dissolved marine sterols are thus an inexhaustible source of ready-made molecules that play a significant role. Dealkylation at C-24 of the C_{29} algal sterols (Fig. 5) has been established for several marine invertebrates (Saliot and Barbier, 1973b; Collignon-Thiennot et al., 1973; Teshima, 1972) and is certainly an adaptive advantage over the usual de novo biosynthesis from acetate. Even for animals or plants which perform total sterol synthesis, the relative amount of sterols in seawater can be a bypass of the system because of the possibility of continuous sterol capture.

In the Atlantic, dissolved cholesterol varies from 12 to 45% (0.24 to 7.30 μg/liter) of total sterols (Saliot and Barbier, 1973a, b), which is an enormous number of tons when it is extrapolated to the mass of the oceans. The average concentration of 3.8 μg/liter contains approximately 6×10^{12}) molecules/ml, which because of its permanent nature is a considerable amount, even for nonplanktonic organisms which continuously capture dissolved substances. 24-Ethylcholesterol (β-sitosterol and its epimer) varies from 32 to 55% of total dissolved sterols, which is significant in connection with the important role of the algal biomass, or, near coasts and estuaries, in relation to continental contribution. Distribution and concentration of dissolved sterols in the oceans, revealing the relative abundance of C_{27} sterols (cholesterol type) and their production by algae, is an unexpected discovery of recent years (Boutry and Jacques, 1970; Ando and Barbier, 1975; Boutry et al., 1971).

C. Peculiarities of Marine Sterols: A Consequence of Adaptation

Some sterols are more abundant in the marine ecosystem than on the continents, even if they do not predominate in the analysis of seawater or of marine organisms. This is the case with 24-methylenecholesterol. Model experiments performed with the marine diatom *Chaetoceros simplex calcitrans* in order to pinpoint the exchanges between the algae and the culture medium (Boutry and Barbier, 1974) have revealed a high rate of production. The diatoms, which contain equal quantities of cholesterol and 24-methylenecholesterol, eliminate the latter sterol so that the final concentration in the medium increases 1400-fold in one week. The reasons for cholesterol retention versus 24-methylenecholesterol elimination are unknown but can be related to cell wall permeability. However, 24-methylenecholesterol is not so abundant in the oceans in spite of its elevated production by phytoplanktonic constituents. This may be connected with its greater sensitivity to oxidation. The accumulation of this sterol in filter-feeding marine invertebrates is well known, in oysters, for example (Idler and Fagerlund, 1957; Salaque et al., 1966), and is due to phytoplankton feeding.

A series of C_{26} sterols have been isolated from many marine animals (Idler et al., 1970; Alcaide et al., 1971; Erdman and Thomson, 1972; Viala et al., 1972). Among these, 24-nor-5,22-cholestadien-3β-ol (24-dimethyl-chola-5,22-dien-3β-ol) (**19**) is present in seawater (Saliot and Barbier, 1973a,b; Tusseau et al., 1978). It is present in phytoplankton (Boutry et al., 1971) and other algae (Ferezou et al., 1974). However, the red alga

Rhodymenia palmata, in which **19** represents 0.45% of its sterols does not incorporate labeled acetate or methionine into **19**. Asterosterol (**20**), the Δ⁷-isomer of **19**, occurs in starfish (Kobayashi *et al.*, 1972) or a tunicate (Erdman and Thomson, 1972; Viala *et al.*, 1972), together with dihydro **21** (Viala *et al.*, 1972; Kobayashi and Mitsuhashi, 1974). The C_{26} homolog of cholesterol (**22**), which was synthesized (Métayer and Barbier, 1972)

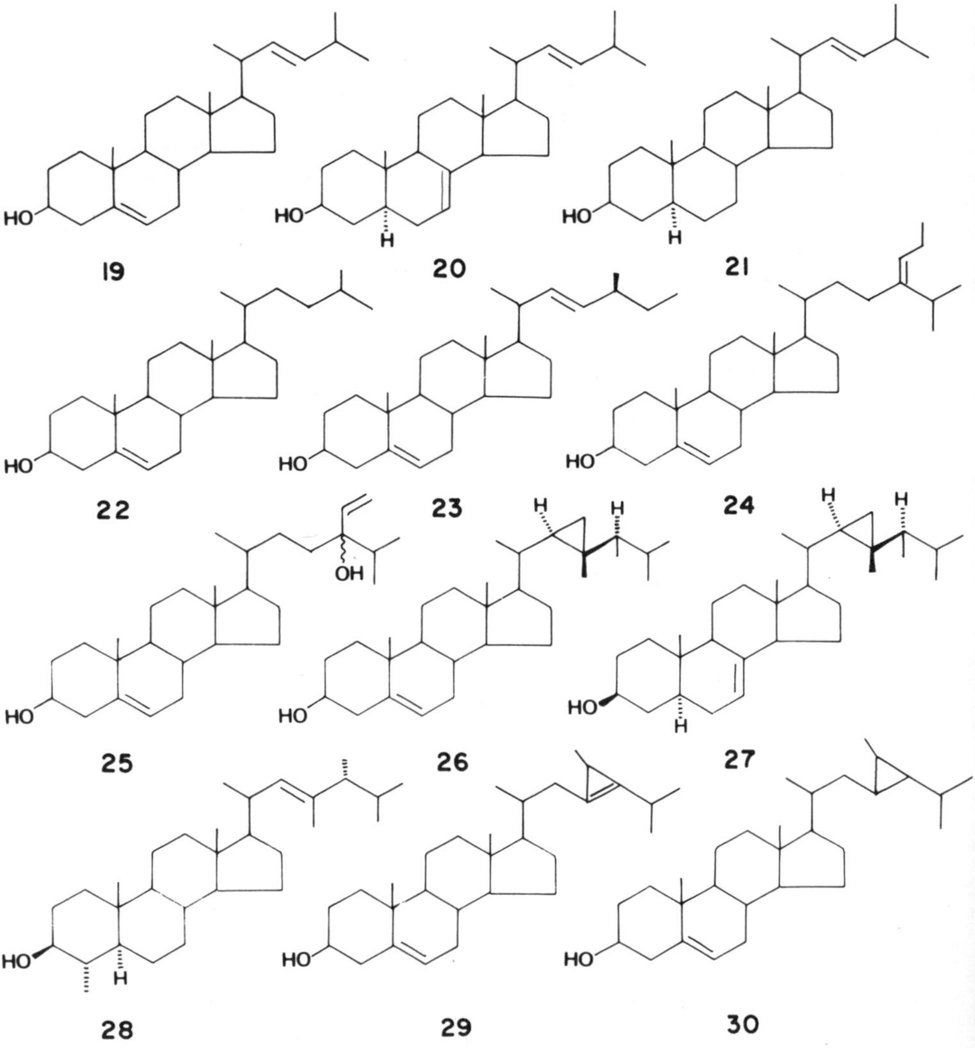

Fig. 6. Some uncommon marine sterols: (**19**) 24-nor-5,22-cholestadien-3β-ol; (**20**) asterosterol; (**21**) 24-nor-5α-cholest-22-en-3β-ol; (**22**) 24-norcholesterol; (**23**) occelasterol; (**24**) 24-propylidenecholesterol; (**25**) saringosterol; (**26**) gorgosterol; (**27**) acanthasterol; (**28**) dinosterol; (**29**) calysterol; (**30**) petrosterol.

before it was reported from a natural source, a scallop (Idler *et al.*, 1976), has recently been found in sponges (Delseth *et al.*, 1979). The presence of C_{26} sterols has been detected in many marine invertebrates, and their possible origin is in plankton production. An interesting, yet to be demonstrated, hypothesis concerns oxidative degradation of C_{28} sterol side chains of the occelasterol type (**23**) (Sheikh and Djerassi, 1975). Astonishingly enough, algal cultures have often been searched for C_{26} sterols (Boutry and Barbier, 1974; Ando and Barbier, 1975), but without success. Triple methylation at C-24 leading to C_{30} sterols (24-propylidene) (**24**) is also to be demonstrated. An intermediate that supports the third biological methylation by S-adenosylmethionine (SAM) remains hypothetical, but saringosterol (**25**) cannot be excluded as a possible candidate, although its status as a natural product is in doubt (Knight, 1970). Unusual side-chain methylations or partial degradations furnish a wide variety of sterols which do not occur in the terrestrial fauna or flora (for a review see Schmitz, 1978). With a combination of both processes, many things may happen to sterol structures; for instance, nor-19-sterols or steroids, and even the corresponding nor-19 C_{26} sterols (Minale and Sodano, 1974) have been detected.

The fact is that such transformations are encountered in many marine invertebrates that shelter colonies of unicellular algae. This is also the case for gorgosterol (**26**), acanthasterol (**27**), or dinosterol (**28**). Recently, Withers *et al.* (1979) succeeded in demonstrating the production of gorgosterol in a culture of the dinoflagellate *Peridinium foliaceum*.

In some sponges which harbor colonies of algae, side-chain degradation gives rise to pregnane and androstane derivatives (Delseth *et al.*, 1978). A similar situation is found in soft corals (Higgs and Faulkner, 1977), which suggests an interaction. The cyclization of fucosterol into calysterol (**29**), a 23,24-cyclopropane sterol related to petrosterol (**30**), has been established in sponges (Minale *et al.*, 1977; Sica and Zollo, 1978).

The question of possible interactions becomes evident in the formation of unusual marine sterols; once again, this must be viewed in terms of adaptation.

Another question which comes to mind is the extent to which the isolation of new marine sterols may be possible. Are the sources of new structures more extensive than was previously believed because of the multiplicity of minor variations, or are they limited by biogenetic aspects? Popov *et al.* (1976) reported the identification of approximately 50 sterols in a single marine organism. The sensitivity of their technique is responsible for such an extensive result. Using a computer-assisted structure manipulation, Varkony *et al.* (1978) predicted the existence of 1778 3-hydroxy natural sterols; as only about 100 are known, most of them remain to be isolated if nature agrees with the computer. The diversity of marine sterols appears to be due to the accumulation of several phenomena, including genetic control, biosynthesis by zooxanthellae, commensalism and parasitism, interactions of all kinds, and transmission of molecules by the food chain or from seawater.

The classical fluid mosaic model of the cell membrane proposes an organized molecular structure endowed with a certain fluidity (Singer and Nicholson, 1972). Sterol esters, often present as liquid crystals (Lee, 1975; Fenton, 1977) within a gel, modulate cell plasticity. The amount of free hydroxysterols in the phospholipid combs of the membrane (Grunwald, 1971; Bean, 1973) controls the circulation of electrolytes or any other mobile cell constituents. Biological methylation of sterol side chains seems to diminish the rate of sterol insertion into the membrane bilayers in the order cholesterol > campesterol > β-sitosterol; unsaturations at C-22, as in stigmasterol or ergosterol, reduce this rate still further (Edwards and Green, 1972). The balance between sodium and potassium ions is greatly impaired by the existence of a C-24 double bond (desmosterol) (Fiehn and Seiler, 1975). In many cases an antagonistic relationship has been noted between the amount of cholesterol in the phospholipid bilayers of the membranes and the efficiency of the enzymatic activity. Hence, sterols appear to have an important function in cell plasticity, membrane permeability, and ion exchanges.

The great variety of marine sterols may be connected with the multiplicity of conditions in the oceans, which introduces the necessity to consider different kinds of adaptation. The exotic permethylation of side chains by unicellular algae may be a variation of the cell–host relationship. Preservation of structures against the host enzymatic systems may be argued if one supposes that gorgosterol (26) or calysterol (29) side chains resist oxidative degradation. This idea needs to be proved. In all this, the need for model experiments in interacting systems is evident, a wide open field for the marine chemist.

The kairomonal role of marine sterols is indeed a special case of chemical ecology, and cholesterol itself appears to be a fundamental molecule of the marine environment.

V. ECOLOGY OF SOME GROWTH FACTORS IN PLANKTON, MACROALGAE, SEDIMENT, AND SEAWATER

A. Vitamin Requirements and Phytoplankton Production

Cultures of diatoms require the addition of several vitamins for normal development. Hutner and Provasoli (1951) and Provasoli and Pintner (1953) have defined the needs of many unicellular marine algae and have rejected the concept of autotrophy for many species. However, things are not that simple: some bacteria and diatoms seem to be able to biosynthesize their own growth factors while others are not. The reasons for this genetic determinism are not evident, but may be rationalized in terms of adaptation once again and may be linked to the ease of finding the necessary factors in the environment. At death, lysis of cells returns to the seawater the compounds which thus, apparently, become the benchmarks for the total planktonic biomass and, by implication, for life on Earth.

Vitamin concentrations in seawater of the order of 1 pg or 1 ng per liter are generally enough to ensure the development of auxotrophic microorganisms and their reproduction. Three growth factors are principally involved and have been studied carefully: vitamins B_{12}, B_1 (thiamine), and H (biotin). Of 179 diatom strains that were investigated, Droop (1962) found 95 unable to develop without the addition of a vitamin mixture. Eighty percent required B_{12}, 53% thiamine, and 10% biotin. A different set of statistics has been reported by Berland et al. (1976) for bacteria; among 232 strains from the Rhone delta (Mediterranean Sea), 8 required B_{12} for their growth, 32 thiamine, and 24 biotin. Bacteria developing on the surface of macroalgae were found (Ericson and Lewis, 1953) to produce B_{12} in 70% of the strains that were studied. Berland et al. (1974a,b) have reviewed the situation of growth factors produced by diatoms and bacteria. The producers of B_{12} are mostly bacteria of the genera *Pseudomonas, Flavobacterium*, and *Achrombacter*. Furthermore, it seems that, on the average, the number of strains producing such vitamins is greater than the number of strains requiring them for their development (Table 2).

As a complication of the ecological aspects of growth factors in the sea, it was shown that while some bacteria are good producers, all of them are avid consumers (Provasoli, 1963). The situation is not simplified by the fact that many microorganisms that do not produce vitamins not only

TABLE 2

Production and Requirements of Growth Factors
by Bacteria[a]

Vitamins	Production (% of strains)	Requirement investigated
B_{12}	26	3.4
B_1 (thiamine)	18	14
H (biotin)	45	10

[a] From Berland et al. (1974a,b).

require them, but actually accumulate them. Provasoli *et al.* (1974) found that of 388 diatom strains, 203 required vitamins, 172 B_{12}, 82 thiamine, and 14 biotin. The concentrating power of diatoms is so marked that they are a source of such vitamins for many macroorganisms and also, unfortunately, good transmitters of pollutants. Bacteria of the sediments (Burkholder, 1959) are often vitamin producers. Of 344 strains investigated, 27% produced B_{12}, 60% thiamine, and 50% biotin.

B. Distribution of Vitamin B_{12}, Thiamine, and Biotin

Seawater is, as one would expect, the ultimate destination of the vitamins, as is demonstrated in a study by Ohwada and Taga (1972) in the northwestern Pacific Ocean, where only 1% of the growth factors was detected in material remaining on filters but 99% in the filtered water. These figures are often modified depending on season or location; for example, coastal waters are poorer in dissolved vitamins and richer in particulates.

Concentrations follow the annual plankton cycles (Carlucci, 1970). In August in the Barentz Sea (Propp, 1970), B_{12} disappears nearly completely from seawater in connection with an intensive development of the auxotroph diatom *Skeletonema costatum,* which is a B_{12} consumer. The disappearance of the diatom biomass in winter, with an augmentation of dissolved organic matter, leads to an increased development of B_{12}-producing bacteria.

Macroalgae have the same concentrating properties as do diatoms. Red algae are known to take B_{12}, which accumulates in the thalli, from seawater or from epiphytic diatoms. *In situ* biosynthesis remains questionable (Provasoli, 1963).

Accumulation of vitamins in marine sediments is sometimes considerable as it arises from bacterial colonies developing on detritic organic matter and from adsorption. Analyses from Puerto Rico indicated the presence of 180 ng/g B_{12} and 73 ng/g thiamine in sediments (Burkholder and Burkholder, 1958).

In view of the large number of known and unknown species together with the facts that until now only diatoms and bacteria have been considered; that other biota such as fungi, protozoans, and metazoans have been neglected; and that a systematic search has been restricted to the three main growth factors, one realizes the enormous task that remains to be accomplished. Collection of pertinent data should have begun with a study of the ocean surface microlayer, which, as mentioned earlier, is the most significant layer of the sea; but, unfortunately, it is also the most difficult to investigate.

C. Growth Factors and Vitamin Content of Marine Macroalgae

Vitamin content of marine macroalgae is subject to important quantitative variation, depending on species, seasons, and locations, all of which may be governed by their ecology. Accumulation of vitamins from seawater or from epiphytic organisms, mainly bacteria and diatoms, established at the algal surface, may interfere with *in vivo* biosynthesis; but this remains to be proved. Average figures, which have been reported by Lundin and Ericson (1956), indicate some differences for B_{12} among brown (0.07 μg/g), red (0.27 μg/g), and green algae (0.35 μg/g dry weight). The limits of variation of vitamin content were reported by Kanazawa (1962, 1963) for 20 species of red algae from Japanese waters: thiamine varies from 0.04 to 4.60 μg/g, and B_{12} from 0.015 to 0.29 μg/g dry weight. The vitamin content of several species of rhodophytes has also been reported. Thiamine content was shown to be 2.8 μg/g in *Chondrus crispus,* 3.04 μg/g in *Rhodomela subfusca,* and 2.82 μg/g in *Porphyra lasciniata* (Gerdes, 1951). Kanazawa (1963) and Kanazawa *et al.* (1966) reported the thiamine content of *Rhodymenia palmata* (6.32 μg/g) and *Grailaria textorii* (4.60 μg/g dry weight). *Rhodymenia palmata* contained 0.18 μg/g of biotin. Other data for riboflavin, nicotinic acid, pantothenic acid, vitamin B_6 complex, and vitamin C in macroalgae have been reviewed by Baslow (1969).

Gibberellins are present in many species (Blunden, 1977) and have also been detected in a commercial algal extract. The use of fresh algae in agriculture is justified by the cytokinin-like substances that are known to accelerate plant growth but whose structures remain to be established. Growth-inhibiting compounds counterbalance (Blunden, 1977) cytokinin activity, but can be removed from seaweed extracts by organic solvents. Healing effects, including some in human clinical cases, have been described for extracts of freshwater algae. This activity could perhaps be related to similar substances (Lefevre, 1964; Lefevre *et al.*, 1963, 1965).

To the author's knowledge, algae are the only plant source of vitamin B_{12} except for its production by bacterial root nodules. This fact was of some interest when algae were eaten fresh as a salad, as is the case with sea lettuce, *Ulva lactuca.* In addition to vitamin B_{12}, this alga contains as much vitamin A as cabbage and as much vitamin C as oranges (Naegelé and Naegelé, 1977); however, it is now completely forgotten as a food. From a general viewpoint, it is sad that the European way of life has excluded the culinary use of algae, which is very different from the traditions of Asia, and particularly Japan. In Japan the shore culture furnishes nearly one million tons of edible seaweed each year, mainly *nori* (*Porphyra,* about 400,000 tons, *kombu* (*Laminaria* sp., about 200,000

TABLE 3

Vitamin Content (per 100 g Dry Weight) of Commercial *Kombu* Samples (*Laminaria* sp.)[a]

Vitamins	Laminaria japonica (Ma kombu)	Laminaria angustata (Noga kombu)	Laminaria ochtensis (Rishili kombu)	Laminaria angustata (Mitsuishi Kombu)	Laminaria religiosa (Hosome kombu)
			Algae species		
Pro-A, IU	430	250	320	360	130
Carotenoids, μg	1300	750	960	1100	390
B$_1$ (thiamine), mg	0.08	0.08	0.06	0.02	0.07
B$_2$, mg	0.32	0.40	0.01	0.20	0.19
Nicotinic acid, mg	1.8	1.8	2	2	3.5
C, mg	11		15		

[a] Prepared in Japan for culinary purposes (Doumenge, 1975).

tons), and *wakame* (*Undaria pinnatifida*). Table 3 (Doumenge, 1975) lists the vitamin content of *kombu,* which is usually found on the market.

It is evident that vitamin B$_{12}$, in contrast with its continental distribution, occupies a privileged bioecological position in the oceans. Until the last century, algal preparations were used in some European kitchens, particularly in England, where *Porphyra laciniata* and *P. vulgaris* were steam-cooked ("laver"), as also were *Ulva latissima* ("green laver") and *Laminaria* sp. ("tangle"). In Ireland, the Scandinavian countries, and on many islands of the North Sea, *Alaria esculenta* was utilized in different ways, for instance, cooked in milk. For a long time flour made from dried algae was mixed with cereal flours to prepare bread (Naegelé and Naegelé, 1977) to which it imparted a special taste. No doubt these practices formerly served as a source of vitamins for the European population.

VI. EXAMPLES OF ECOTOXICOLOGY OF THE OCEANS

A. The Mercury Cycle

The Whittaker and Feeny (1971) classification of natural interactions places pollution at two levels: with the allomones (depressants), and among the intraspecific effects (when a toxin acts on its producer). It does not seem possible to include mineral elements, as they are not produced by an excreting organism, except after transformation into an organic derivative. This is the case with mercury, for which no biochemical function seems to have been established, but which enters into bacterial enzyme systems and is further metabolized into mono- and dimethylmercury. This reaction, which should be SAM-dependent, has apparently no function other than elimination.

The natural cycle of mercury depends on a series of factors: washing of the continents by rain; the action of volcanoes; the population of marine sediment bacteria; and the concentrating power of plankton. Bacteria in the sediments (for a review see Ramade, 1978) transform mineral mercury into monomethylmercury and the volatile dimethylmercury $(CH_3)_2Hg$. Monomethylmercury, however, stays in the ocean, where it is captured by phytoplankton. There exists therefore, via the atmosphere and vertebrate animals (birds and man), a continuous exchange of this element. Normal mercury content in the atmosphere is 0.002 ppb and less than 1 ppb in water, rivers, and oceans, which is a satisfactory dilution. But due to the concentrating properties of phytoplankton, the mercury level in carnivorous fish reaches above 100 ppb, without any pollution. Tuna naturally contain on the average 120 ppb of mercury (Ramade, 1978), and high concentrations have also been detected in centuries-old skeletons found in museums, and even in 10-million-year-old fossils, findings which document the permanence of natural mercury pollution. A cycle is represented in Fig. 7, and details of an ecosystem analyzed in a Swedish lake (Westöö, 1966) are presented in Table 4. The question of whether a given molecule has no biochemical significance should perhaps be answered in the affirmative; but the question of whether a given molecule provides no biochemical interference, should be answered negatively. This indeed

TABLE 4

Naturally Occurring Concentration of Mercury as Determined in a Swedish Lake[a]

Source	Concentration (ppb)
Water	0.1
Phytoplankton	10–100
Zooplankton	100–500
Aquatic insects	0.1–1
Microphagous fish	0.5–1
Carnivorous fish (pike)	4
Other values[b]	
Pike	0.1 ppm
In a polluted area	10 ppm

[a] From Westöö (1966).
[b] Values taken from Ramade (1978).

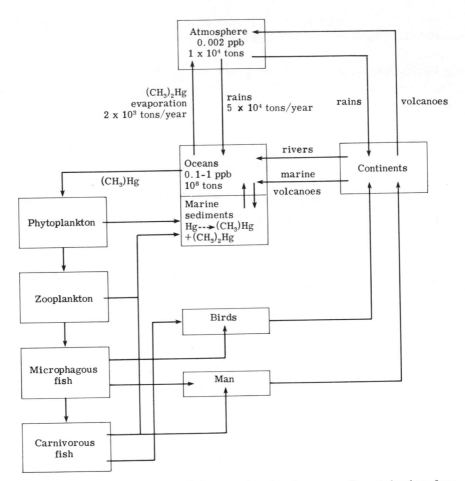

Fig. 7. A simplified scheme of the natural cycle of mercury. Reported values from Ramade (1978) and Sivick *et al.* (1974).

appears to be the case for mercury; but the two systems overlap somewhere, and a firm conclusion remains difficult.

World industrial production of mercury doubled in the 25 years between 1946 and 1971 (Ramade, 1978), and growth studies indicate the exponential character of the curve. Production was 10,000 tons in 1971 (Goldwater, 1971), and the amount released to the atmosphere by the use of coal and other fuel was estimated at 3,000 tons (Jeensu, 1971). But this last figure is surely an underestimate because it concerns only coal, and uses 1 ppm of mercury in coal as the basis for calculation, but the average value is 3.3 ppm. The author rationalizes his preference as a "conservative estimate.'' In fact, more than 10,000 tons of mercury are introduced each year into the atmosphere by the combustion of fuel and coal. The total amount of mercury ''handled'' by man should be around 20,000 tons at present, but this is of course an approximation. Twenty-six percent of the mercury produced is employed in the preparation of fungicides and other pesticides, 33% in electrochemistry for sodium and chlorine production, and 41% in the electrical industry and in laboratory uses; but this last value should be revised upward due to the current considerable need for mercury in electronics.

Transfer of continental mercury to the oceans by rivers has been estimated at 5000 tons/year (Goldberg, 1970). The sediments of the St. Clair river in Canada, prior to establishment of strict controls, contained 1700 mg of mercury per kilogram (Grimstone, 1972).

The dose that leads to possible nervous disorders in man is 6 ppm, and the permitted concentration is 0.5 ppm, but these values need to be reconsidered because of possible accumulation. The period of elimination in man is slow, 70 days for 50%; moreover, methylmercury is stable and passes through the placenta. The danger of methylmercury in the oceans is of course linked to the accumulation phenomenon and to the necessity of bacterial interaction for transformation to the volatile dimethylmercury. Numerous diseases and deaths have demonstrated the real danger of mercury pollution as in the cases of Minamata and Niigata, Japan. In these particular instances, mercuric chloride was disposed as waste in the manufacture of vinyl chloride and acetaldehyde. According to Ui (1971) and Ramade (1978), some Japanese scientists were offered money in order to show the impossibility that mercury was involved in the Minamata affair. However, ecological data have demonstrated the true relationships. According to these data, the mercury content of seawater in Minamata Bay was only 0.1 ppb, but 50 ppm was found in fish (a 500,000-fold concentration). The river Agano flowing into the bay contained 0.1 μg/l methylmercury; diatoms 10 ppm, and fish 40 ppm (400,000 times the initial concentration).

Mercury contamination is strictly controlled everywhere now, but the problem has become larger, if less visible, because of increased production and the ease of concentration accumulation. The regular consumption of fish and other seafood may become questionable and will certainly be a problem for tomorrow. Closed systems are of course more vulnerable, and the Mediterranean Sea, with its highly industrialized areas, is in particular danger. Results of mercury analyses in fish, mussels, and man on the French Mediterranean coast are reported in Table 5.

Like marine animals, macroalgae are also able to concentrate mercury from seawater, which was shown through systematic research using an

TABLE 5

Mercury Determination on the French Mediterranean Coast[a]

Samples	Total mercury (ppm/dry weight)	Methylmercury (ppm/dry weight)
Sea perch, *Serranus* sp.	0.99	0.43
Mussel, *Mytilus edulis*	2.58	0.06
Sardine, *Arengus minor*	0.46	0.22
Whiting, *Merlangus merlangus*	0.87	0.58
Fisherman No. 1	7.39	7.16
Fisherman No. 2	1.58	0.75
Research worker No. 1	3.66	2.58
Research worker No. 2	2.88	1.59

[a] From Ui (1971); Ramade (1978).

atomic absorption spectrometer and comparing the accumulative properties of seven species (Sivick *et al.*, 1974). The total amount varies between
0.03 ppm for *Dictyopteris undulata* and 0.3 ppm for *Sargassum
johnstonii*. Most values are above the 0.05-ppm limit now fixed by the
World Health Organization. Mercury was found in carrageenan, proteins,
and lipids. The question remains, in what form mercury is present in these
algae.

No one can really tell what will occur in the near future with the
development of inevitable and progressive intoxication. The dystrophy of
the system could mark the end of a cycle and the beginning of an unexpected unnatural condition. The alarm has been sounded for years (Club
de Rome, 1972, 1974; Peccei, 1975; Toffler, 1974; Barbier, 1976) on the
dangers inherent to the evolution of human society, and no solution is in
sight. The case of the mercury cycle is just one example.

B. Hydrocarbon Spillage and Its Effect on the Biosphere

Life probably began in the water, is water-dependent, and the forms of
life that have a conscience should respect water. Most specialists who
regularly observe the sea bottom agree with the conclusion that speciation
and density of life are regressing (Cousteau, 1970). This is the result of
pollution, mainly from the heavy metals mercury and lead, from pesticides including halocarbons, from detergents, and most significantly
from hydrocarbons. About 10 years ago, hydrocarbon pollution was estimated at some 5 million tons a year (ZoBell, 1964; Pilpel, 1968). The
Torrey Canyon story, when 100,000 tons of oil spilled onto the Seven
Stones Bank, southwest of England, is now out of date. The observed

catastrophic effects on the marine flora and fauna have been repeated so many times since that a catalog of these accidents has become necessary.

Natural processes of hydrocarbon elimination in the oceans have been observed. The lightest fractions evaporate; other hydrocarbons are fractionated, dispersed by the tides and waves, and air-oxidized; still others drift to the bottom where they are destroyed by bacteria. Extensive use of adsorbents, dispersing agents, and detergents tends to exert long-term effects on the marine environment. The immediate effect is unavoidable, particularly on defenseless birds which disappear in great numbers when large surfaces are covered. The action of ultraviolet irradiation on oil constituents leads to the formation of aliphatic hydroperoxides, followed by fragmentation into smaller molecules which are more easily assimilated (Pilpel, 1968; Nixon, 1972; Dzuban, 1958). Drifting to the sea bottom occurs very slowly, sometimes over several months. Microbial attack begins instantly because of bacteria–protozoa equilibria. Unfortunately, chemicals, particularly detergents, counteract this natural process (George, 1961).

Hydrocarbons are natural constituents of seawater (Barbier et al., 1973). Interesting deductions can be made when one analyzes ancient waters, with a turnover taking place over centuries, prior to heavy pollution. Natural hydrocarbons dissolved in seawater are rapidly extracted by marine invertebrates and metabolized (Fevrier et al., 1975). Hydrocarbons are thus included in a series of cycles, which lead to their elimination (Fig. 8). But this does not mean that hydrocarbon pollution is not a real problem. While petroleum reserves are being depleted everywhere in the world, new fields are also being discovered (e.g., the Mexican field), and further deep-sea drilling must be expected. In England, 250,000 birds are disappearing each year due to oil pollution. In Newfoundland (Dorst, 1965) about 200,000 penguins have disappeared over 15 years. Some isolated, now forgotten, incidents have led to fantastic effects. For example, the wreck of the tanker "Gerd Maersk" in the Ebre river, Spain 1955, was responsible for the destruction of 500,000 velvet scoter ducks, Melanitta fusca (Ramade, 1978).

The formation of solid spherical particles after evaporation of the light-petroleum fractions and dispersion of the oil layer has curious ecological consequences which must be mentioned. The spheres vary from about 0.1 to 10 cm in diameter and are covered by bacteria. They float and support colonies of the isopod Idotea metallica and of the barnacles Lepas anatifa and L. fasciculatus, which have thus found a new means of transport (Horn et al., 1970). A statistical study of the sphere formation yielded values of 1 mg/m² for the Atlantic Ocean and 20 mg/m² for the Mediterranean Sea (Morris, 1971). The final conclusion based on

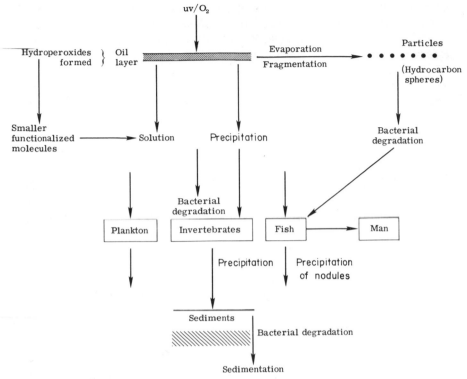

Fig. 8. The natural fate of oil-spill hydrocarbons in the oceans and their eventual elimination.

this hydrocarbon particle count, is an estimate of 760,000 tons of petroleum annually for the Atlantic and 530,000 tons for the Mediterranean (1971 data).

The smallest hydrocarbon particles are directly transmitted to man via the fish market. Local situations may be more dramatic than indicated above. For example, in the Sargasso Sea and around Bermuda, because of the conjunction of the Gulf Stream, the North Equatorial Current, and the Canary Current, a constant accumulation of particles, lumps, and all kinds of petroleum residues occurs (Butler *et al.*, 1973). The concentration of particles and lumps in the Sargasso Sea is between 2 and 40 mg/m^2 (about 64,000 tons), while the beach deposits are of the order of 4 to 1700 g/m along the Bermuda shoreline.

C. Pollution and Biological Concentrators

Some marine organisms, as already mentioned, are good biological concentrators and may thus modify the profile of any trophic relationship.

Hydrocarbons accumulate easily in invertebrates (Février *et al.*, 1975) and diatoms (Boutry *et al.*, 1977). Inorganics are often concentrated, which has been known for a long time since iodine is concentrated by the algae *Laminaria* sp. and *Fucus* sp. which are used for extraction of iodine. The vanadium level is relatively high in the blood of squids (Ramade, 1978) and the hepatopancreas of *Pecten maximus* contains 0.1% cadmium (dry weight). Freshwater phytoplankton concentrate radiophosphorus (Foster and Rostenbach, 1954) to the extent that it contains 1000 times the initial amount found in water. The same phenomena occur in the oceans with chlorinated insecticides, which are often undetectable in water but may be found in the phytoplankton. This phenomenon cannot be ignored by the marine chemist and has many consequences which need to be examined seriously.

In the North Sea the dieldrin (**31**) content in water is undetectable, but it is 1 ppb in the phytoplankton. Oysters contain several thousand times this value, 70,000 times in the case of *Crassostrea virginica* (Butler, 1965).

The introduction of concentrators into the trophic chains explains why pisciverous birds collect trace elements from everywhere and to a greater extent than any other organism (George and Frear, 1966). Among these, cormorant birds, which are exclusive fishers, accumulate more than do birds that eat fish, crustaceans, and mollusks (e.g., sea gulls).

A study of polychlorobiphenyl (PCB)-enriched slicks of the Sargasso Sea revealed that the concentration in the first 150-μm layer was 11.2 μg/l, while at 30-cm depth it was 3.6 μg/l (Bidleman and Olney, 1974). The concentration of the chlorinated products in the surface microlayer influences the development of the plankton. The effects of chlorinated hydrocarbons, polybiphenyls, etc. on phytoplankton have been studied (Wurster, 1968). Inhibition of photosynthesis could be observed beginning with a level of 10 ppb (Chasseaud, 1970). As already mentioned, in the Minamata mercury pollution in Japan, an enrichment of about 500,000-fold was determined in fish caught near the factory with the mercury effluent.

The final destiny of any product accumulated by an ecological concentrator is the ocean bottom, as seen with DDT (Woodwell *et al.*, 1971). Dieldrin (**31**) concentration in marine organisms can be measured all along the food chain, as shown in Table 6 (Ramade, 1978).

The existence of selective concentrating power in the marine biosphere should introduce caution and care among chemists who are interested in the chemistry of marine organisms. The exact origin of a compound cannot be ascertained until biosynthetic experiments that exclude interactions and the possible role of symbiotic flora are performed. From another viewpoint, the notion of a biological marker can be retained and become a useful tool for the study of trophic chains in the marine environment.

TABLE 6

Dieldrin (31)[a] Concentration in Some Marine Organisms[b]

Source	Amount (ppm)
Seawater	Undetectable
Phytoplankton	1×10^{-3}
Zooplankton	2×10^{-2}
Crustaceans and	
microphagous fish	3×10^{-2}
Carnivorous fish	0.2
Polyphagous birds	
Sea gull	0.1
Stern	0.2
Cormorant birds	
Liver	1.6
Eggs	1.2

[a] Dieldrin (31) (HEOD): 1,2,3,4,10,10-hexachloro - 6,7 - epoxy - 1,4,4a,5,6,7,8,8a - octahydro-*endo*-1, 4-*exo*-5,8-dimethanonaphthalene and alternative representation:

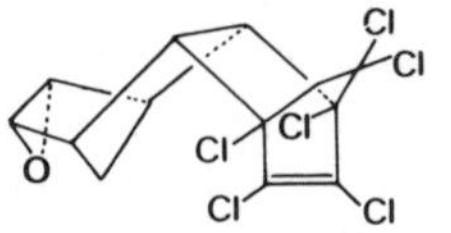

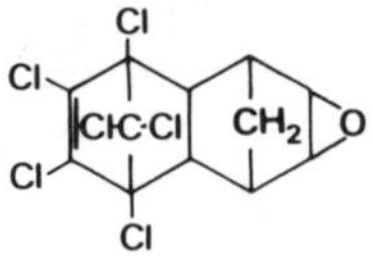

[b] Ramade (1978).

The concentrating power of marine organisms has probably reached its highest values for chemicals from plastics factories. Vinyl chloride, butyl and octyl phthalates are present everywhere in seawater (Copin and Barbier, 1971). Phthalates have been found at a concentration of 3.2 mg/kg in fish (Mayer *et al.*, 1972). The collection in some places of plastic bottles, lumps, sheets, and balls, produced by the action of waves and currents, on the ocean bottom gives an idea of this invasion (Colton *et al.*, 1974). Fish mortality due to ingestion of such plastic balls or stoppers is increasing continually.

However, of crucial importance in the near future will certainly be the accumulation of radioactive elements in the concentrators—and the production of wastes from atomic centers is supposed to reach 10 million tons a year by 2000 (Colas, 1973). This no doubt will be the most serious problem to be considered by the marine ecologist of tomorrow. The accumulation of pollutants in concentrators must be carefully monitored for all commercial species.

VII. CONCLUDING REMARKS

The term ecology has been considerably misused in recent years by introducing a kind of fantasy into the serious problems related to life, the relationships among living organisms, and particularly, the rapid development of human society. Chemical ecology represents a first attempt to classify, understand, and treat natural phenomena at the molecular level. Each compound and its significance are placed into a system, where it plays a role as a messenger, a behavioral instrument, or a biochemical tool. Marine chemical ecology is based on a relatively isolated system, as the oceans are connected to the continents through a series of cycles, and where the surface ultramicrolayer plays the most determining role. One of the main topics in these pages has been to point out the particulars of such a giant ecosystem. First, vital molecules such as sterols and vitamins were considered as examples together with some important phenomena, symbiosis, antibiosis, and the concentrating power of phytoplankton.

Pollution is not really man-made. Natural pollution such as the intoxication of the marine environment by terrestrial and telluric mercury has existed at all times. Man, however, has accelerated the process, which thus leads to a dangerous situation.

This is not the place to discuss in detail the prospective work of the Club de Rome, which uses computers to study the near future by choosing pollution as one of the five main factors that condition the planet. It should be noted, however, that their conclusions are very pessimistic, and the scenarios that emerge from the machines suggest several dramatic possibilities in the next century. The Haldane proposals, according to which man has no chance to realize future equilibrium on Earth, except through a preliminary destruction of his society or the rapid establishment of a powerful world organization, are very pessimistic. Such a defeatist conclusion is opposed to awareness of the problem and the idea that something should be done rapidly.

By turning to a study of natural phenomena, we should, rather, remain optimistic, with marine chemical ecology as a gateway to discovery of new biologically active molecules in a world where numerous interactions lead to the unsuspected. In order to accomplish this, it will be necessary to abandon systematic search for a given biological activity, which of course cannot lead to anything unexpected.

The predictive value of marine chemical ecology, and of chemical ecology as a whole, is still feeble, but when used with other data may be significant. Not only is it a new field open to new possibilities for the chemist interested in natural substances, but it is also a way toward a better understanding of nature. The main targets of marine chemical ecology are these: (1) an attempt to understand and classify the natural phenomena according to molecular determinism; (2) a source of new and unsuspected biological activities and a key to the search for the molecules responsible for the activity; and (3) a comprehensive approach to the study of the equilibria that govern life on the planet.

Research in marine chemical ecology needs, of course, interdisciplinary collaboration, which often poses a problem; it must also be started with simple phenomena based on observations of nature.

REFERENCES

Abeloos, M., and Teissier, G. (1926). *Bull. Soc. Zool. Fr.* **51**, 145.

Abraham, E. P. (1967). *Q. Rev.* **21**, 231.

Alcaide, A., Viala, J., Pinte, F., Itoh, M., Nomura, T., and Barbier, M. (1971). *C. R. Acad. Sci. Paris Ser. C* **273**, 1396.

Allais, J. P., and Barbier, M. (1977). *FEBS Lett.* **82**, 333.

Allais, J. P., Alcaide, A., and Barbier, M. (1973). *Experientia* **29**, 944.

Ando, T., and Barbier, M. (1975). *Biochem. System. Ecol.* **3**, 245.

Aubert, M. (1976). Actualités de Biochimie Marine,'' Colloque Gabim, p. 179. CNRS, Paris.

Aubert, M., Pesando, D., and Pincemin, J. M. (1972). *Rev. Int. Oceanogr. Med.* **25**, 17.

Awata, N., Morisaki, M., and Ikekawa, N. (1975). *Biochem. Biophys. Res. Commun.* **44**, 157.

Barbier, M. (1966). *Ann. Abeille* **9**, 243.

Barbier, M. (1976). ''Introduction à l'Ecologie Chimique,'' p. 119. Masson, Paris; (1978). *ibid.*, rev. ed., p. 229. in Russian. Mir, ed., Moscow; (1979). ''Introduction to Chemical Ecology,'' p. 128. Longman, London.

Barbier, M., and Saliot, A. (1976). ''Actualités de Biochimie Marine,'' Colloque Gabim, p. 63. CNRS, Paris.

Barbier, M., Joly, D., Saliot, A., and Toures, D. (1973). *Deep-sea Res.* **20**, 236.

Baslow, M. H. (1969). ''Marine Pharmacology,'' (Wilkins, Ed.), p. 285. Krieger, New York.

Bean, G. A. (1973). *Adv. Lipids Res.* **11**, 193.

Bergmann, W., and Feeny, R. J. (1950). *J. Am. Chem. Soc.* **72**, 2809.

Bergmann, W., and Feeny, R. J. (1951). *J. Org. Chem.* **16**, 981.

Berland, B. R., Bonin, D. J., Fiala, M., and Maestrini, S. Y. (1947a). ''Substances Naturelles dissoutes dan l'eau de mer,'' p. 121. CNRS, Paris.

Berland, B. R., Bonin, D. J., and Maestrini, S. Y. (1974b). Thesis, Marseille Univ., France, p. 239.

Berland, B. R., Bonin, D. J., Durbec, J. P., and Maestrini, S. Y. (1976). *Hydrobiologia* **50**, 167.

Bidleman, T. F., and Olney, C. E. (1974). *Science* **183**, 516.

Bloch, A. (1975). *Ann. N.Y. Acad. Sci.* **255**, 1.

Blunden, G., in Faulkner, D. J., and Fenical, W. H. (1977). ''Marine Natural Products Chemistry,'' p. 337. Plenum Press, New York.

Boutry, J. L., and Barbier, M. (1974). *Mar. Chem.* **2**, 217.

Boutry, J. L., and Jacques, G. (1970). *Bull. Soc. Chim. Biol.* **52**, 348.

Boutry, J. L., Alcaide, A., and Barbier, M. (1971). *C. R. Acad. Sci. Paris* **272**, 1022.

Boutry, J. L., Bordes, M., Février, A., and Barbier, M. (1977). *J. Exp. Mar. Biol. Ecol.* **30**, 277.

Burkholder, P. R. (1959). *Int. Oceanogr. Congr., Am. Assoc. Adv. Sci., New York* p. 1022.

Burkholder, P. R., and Burkholder, L. M. (1958). *Bull. Mar. Sci. Gulf. Caribbean* **8**, 201.

Butler, P. A. (1965). *U.S. Fish Wildlife Circ.* **226**, 65.

Butler, J. N., Morris, B. F., and Sass, J. (1973). Pelagic tar from Bermuda and the Sargasso Sea, *Bermuda Biol. Station Rep.* **10**, 346.

Carlucci, A. F. (1970). *Bull. Scripps Inst. Oceanogr.* **17**, 103.

Chasseaud, L. F. (1970). Foreign compound metabolism in mammals, *Chem. Soc. London* **1**, 75.

Chen, S. M. L., Nakanishi, K., Awata, N., Morisaki, M., Ikekawa, N., and Shimizu, Y. (1975). *J. Am. Chem. Soc.* **97**, 5297.

Ciereszko, L. S. (1976). *In* "Marine Natural Products Chemistry." (D. J. Faulkner and W. H. Fenical, eds.). Plenum Press, New York.

Club de Rome (1972). "Halte à la Croissance," p. 210. Fayard, Paris.

Club de Rome (1974). "Stratégie pour Demain," p. 185. Le Seuil, Paris.

Codomier, L., Bruneau, Y., Combeau, G., and Teste, J. (1977). *C. R. Acad. Sci. Paris Ser.* D **284**, 1163.

Colas, R. (1973). "La Pollution des Eaux," p. 126. Que Sais-je ed., Presse Univ. de France, Paris.

Collignon-Thiennot, F., Allais, J. P., and Barbier, M. (1973). *Biochimie* **55**, 579.

Colton, J. B., Knapp, F. D., and Burns, B. R. (1974). *Science* **185**, 491.

Copin, G., and Barbier, M. (1971). *Cahiers Oceanogr.* No. 5, 455.

Cousteau, C. (1970). Information du Conseil de L'Europe.

Delseth, D., Carlson, R. M. K., Djerassi, C., Erdman, T. R., and Scheuer, P. J. (1978). *Helv. Chim. Acta* **61**, 1470.

Delseth, C., Tolela, I.., Scheuer, P. J., Wells, R. J., and Djerassi, C. (1979). *Helv. Chim. Acta* **62**, 101.

Devys, M., and Barbier, M. (1967). *Bull. Soc. Chim. Biol.* **49**, 865.

Devys, M., Alcaide, A., Pinte, F., and Barbier, M. (1969a). *C. R. Acad. Sci. Paris* **269**, 2033.

Devys, M., Alcaide, A., and Barbier, M. (1969b). *Phytochemistry* **8**, 1441.

Dorst, J. (1965). "La Nature Dénaturée," p. 190. Delachaux, Paris; "Avant que Nature ne Meure," p. 424. Delachaux, Paris.

Doumenge, F. (1975). *Bull. Soc. Languedoci. Géog.* **9**, 119.

Doyle, P. Y., Patterson, G. W., Dutky, S. R., and Thomson, M. Y. (1972). *Phytochemistry* **11**, 1951.

Droop, M. R. (1962). *In* "Physiology and Biochemistry of Algae," (R. A. Lewin, ed.), p. 141. Academic Press, New York.

Dzuban, I. N. (1958). *Byull. Inst. Biol. Vodokhan Nauk USSR* **1**, 11.

Edwards, P. A., and Green, C. (1972). *FEBS Lett.* **20**, 97.

Eisner, T. (1972). *Verhandl. Deut. Zool. Ges.* **65**, 123.

Erdman, T. R., and Thomson, R. H. (1972). *Tetrahedron* **28**, 5163.

Ericson, L. E., and Lewis, L. (1953). *Ark. Chem.* **6**, 247.

Faulkner, D. J. (1977). *Tetrahedron* **33**, 1421.

Faulkner, D. J. (1979) in Sammes, P., "Topics in Antibiotic Chemistry" (P. Sammes, ed.), p. 13. Elliswood Publ., New York.

Fenton, D. E. (1977). *Chem. Soc. Rev.* **6**, 325.

Férezou, J. P., Devys, M., Allais, J. P., and Barbier, M. (1974). *Phytochemistry* **13**, 595.

Février, A., Barbier, M., and Saliot, A. (1975). *C. R. Acad. Sci. Paris Ser.* D **281**, 239.

Février, A., Barbier, M., and Saliot, A. (1976). *J. Exp. Mar. Biol. Ecol.* **25**, 123.

Fiehn, W., and Seiler, S. (1975). *Experientia* **31**, 773.

Fontaine, M. (1976). "Actualités de Biochimie Marine," p. 5. Colloque Gabim, CNRS, Paris.

Foster, R. F., and Rostenbach, R. E. (1954). *J. Am. Water W.K.S. Assoc.* **46**, 640.

Gagosian, R. B. (1975). *Deep-Sea Res.* **39**, 1443.

George, M. (1961). *Nature (London)* **192,** 1209.

George, J. L., and Frear, D. E. H. (1966). *J. Appl. Ecol. Suppl.* **3,** 155.

Gerdes, G. (1951). *Arch. Mikrobiol.* **16,** 53.

Glombitza, K. W. (1970). *Planta Medica* **18,** 210.

Glombitza, K. W. (1977). *In* "Marine Natural Products Chemistry" (D. J. Faulkner and W. H. Fenical, eds.), p. 191. Plenum Press, New York.

Glombitza, K. W., Rösener, H. V., Vilter, H., and Rawald, W. (1973). *Planta Medica* **24,** 301.

Goad, L. J. (1978). *In* "Marine Natural Products" (P. J. Scheuer, ed.), Vol. 2, p. 76. Academic Press, New York.

Goldberg, E. (1970). *In* "Global Effects of Environmental Pollution," p. 178. Reidel, Dorschecht.

Goldwater, L. J. (1971). *Sci. Am.* **224,** 15.

Grant, P. T., and Mackie, A. M. (1977). *Nature (London)* **267,** 786.

Grimstone, G. (1972). *Chem. Br.* **8,** 244.

Grünwald, C. (1971). *Plant Physiol.* **48,** 653.

Harborne, J. B. (1977). "Introduction to Ecological Biochemistry," p. 243. Academic Press, New York.

Higa, T., and Scheuer, P. J. (1975). *Tetrahedron* **31,** 2379.

Higgs, M. D., and Faulkner, D. J. (1977). *Steroids* **30,** 379.

Horn, M. H., Teal, J. M., and Backus, R. H. (1970). *Science* **168,** 245.

Hügel, M. F., Barbier, M., and Lederer, E. (1964). *Bull. Soc. Chim.* 2012.

Hutner, S. H., and Provasoli, L. (1951). *Biochem. Physiol. Protozoa* 27.

Idler, D. R., and Fagerlund, U. H. M. (1957). *J. Am. Chem. Soc.* **79,** 1988.

Idler, D. R., Wiseman, P. M., and Safe, L. M. (1970). *Steroids* **16,** 451.

Idler, D. R., Khalil, M. W., Gilbert, J. D., and Brooks, C. J. W. (1976). *Steroids* **27,** 155.

Jaenicke, L. and Müller, D. G. (1973). *Fortschr. Chem. Org. Naturstoffe* **30,** 61.

Jeensu, O. I. (1971). *Science* **172,** 1027.

Jensen, A., Rystad, B., and Skoglund, L. (1972). *J. Exp. Mar. Biol. Ecol.* **8,** 241.

Kanazawa, A. (1962). *Mem. Fac. Fish Kagoshima Univ.* **10,** 31.

Kanazawa, A. (1963). *Nippon Suisan Gukhaishi* **29,** 713.

Kanazawa, A., and Teshima, S. (1971). *J. Oceanogr. Soc. Jpn.* **27,** 207.

Kanazawa, A., and Teshima, S. (1978). *Oceanolog. Acta* **1,** 39.

Kanazawa, A., Saito, A., and Idler, D. R. (1966). *J. Fish. Res. Bd. Can. J.* **23,** 915.

Kazlauskas, R., Murphy, P. T., Quin, R. J., and Wells, R. J. (1977). *Tetrahedron Lett.* 37.

Kirschenblatt, J. D. (1957). *Trav. Soc. Natur. Leningrad* **73,** No. 4, 225.

Kirschenblatt, J. D. (1962). *Nature (London)* **195,** 916.

Kitagawa, I., Nishino, T., and Kyogoku, Y. (1979). *Tetrahedron Lett.* 1419.

Knight, B. A. (1970). *Phytochemistry* **9,** 903.

Kobayashi, M., and Mitsuhashi, H. (1974). *Steroids* **24,** 399.

Kobayashi, M., Tsuru, R., Todo, K., and Mitsuhashi, H. (1972). *Tetrahedron Lett.* 2935.

Lavoisier, A. L., de (1792). *In* "Les Classiques du Peuple" (E. Kahane, ed.), p. 271. Lavoisier, Paris.

Law, J. H., and Regnier, F. E. (1971). *Ann. Rev. Biochem.* **40,** 533.

Lederer, E. (1969). *Q. Rev.* **23,** 453.

Lederer, E., Teissier, G., and Huttrer, C. (1940). *Bull. Soc. Chim. Fr.* **7,** 603.

Lee, A. G. (1975). *Endeavour Fr.* **34,** 67.

Lefevre, M. (1964). *In* "Algae and Man," p. 337. Plenum Press, New York.

Lefevre, M., Laporte, G., and Flandre, O. (1963). *C. R. Acad. Sci. Paris* **256,** 254.

Lefevre, M., Laporte, G., and Flandre, O. (1965). *In* "La Cicatrisation," p. 217. CNRS, Paris.

Li, C. P., Goldin, A., and Hartwell, J. L. (1974). Cancer Chemotherapy Rep., **4,** Part 2, 97.

Lovell, F. M. (1966). *J. Am. Chem. Soc.* **88,** 4510.

Lovelock, J. E. (1975). *Nature (London)* **256,** 193.

Lundin, H., and Ericson, L. E. (1956). *Int. Seaweed Symp.* p. 39. Pergamon, Oxford.

McConnell, O. J., and Fenical, W. (1977). *Tetrahedron Lett.* 1851.

MacIntyre, F. (1975). "The Sea" (E. D. Goldberg, ed.), p. 245. Wiley, New York.

Mayer, F. L., Stalling, D. L., and Johnson, J. L. (1972). *Nature (London)* **238,** 411.

Métayer, A., and Barbier, M. (1972). *Bull. Soc. Chim. Fr.* 3625.

Minale, L., and Sodano, G. (1974). *J. Chem. Soc. Perkin I* 1888.

Minale, L. *et al.* (1977). *Experientia* **33,** 1550.

Moody, K., Thomson, R. H., Fattorusso, E., Minale, L., and Sodano, G. (1972). *J. Chem. Soc. Perkin I* 18.

Morris, B. F. (1971). *Science* **173,** 430.

Naegelé, E., and Naegelé, A. (1977). "Les Algues," p. 97. Presses Univ. France, Coll. Que Sais-je.

Nes, W. R., and MacKean, M. L. (1977). "Biochemistry of Steroids and Other Isopentenoids," p. 690. Univ. Park Press, London.

Nixon, A. C. (1972). "Autoxidation et Antioxydants," Vol. II, p. 695. Wiley (Interscience), New York.

Ohta, K. (1977). *Agr. Biol. Chem. Jpn.* **41,** 2105.

Ohta, K., and Tagagi, M. (1977). *Phytochemistry* **16,** 1085.

Ohwada, K., and Taga, N. (1972). *Mar. Chem.* **1,** 61.

Oldham, G., Norèn, B., Norkrans, B., Södergren, A., and Löfgren, H. (1978). *Progr. Chem. Fats Other Lipids* **16,** 31.

Pasteur, L. (1887). Asselineau, J., and Zalta, J. P. (1973). *In* "Les Antibiotiques," p. 364. Hermann, Paris.

Peccei, A. (1975). *L'heure de la Vérité,* Fayard ed., Paris, 210.

Pilpel, N. (1968). *Endeavour ed. Fr.* **27,** 11.

Popov, S., Carlson, R. M. K., Wegmann, A., and Djerassi, C. (1976). *Steroids* **28,** 699.

Premuzic, E. (1971). *Fortschr. Chem. Org. Naturstoffe* **29,** 417.

Privat de Garilhe, M. (1977). *Actual. Chimi.* 20.

Privat de Garilhe, M., and Rudder, D. (1970). *Tokyo Univ. Press* **2,** 180.

Propp, L. N. (1970). *Okeanologiya,* **10,** 851.

Provasoli, L. (1963). "The Sea" (M. N. Hill, ed.), p. 165. Wiley (Interscience), New York.

Provasoli, L., and Pintner, I. J. (1953). *N.Y. Acad. Sci.* **56,** 839.

Provasoli, L., Carlucci, A. F., and Steward, W. D. P. (1974). Algal Physiology Biochemistry, *In* "Botanical Monographs," Vol. 10, p. 741. Blackwell, Oxford, 741.

Ramade, F. (1978). "Eléments d'Ecologie Appliquée," p. 522. McGraw-Hill, Paris.

Raper, J. R. (1952). *Botan. Rev.* **18,** 447.

Reschke, T. (1969). *Tetrahedron Lett.* 3435.

Rinehart, K. C. *et al.* (1976). *Phys. Chem. Sci. Res. Rep.* **1,** 651.

Rohmer, M., and Brandt, D. R. (1973). *Eur. J. Biochem.* **36,** 446.

Salaque, A., Barbier, M., and Lederer, E. (1966). *Comp. Biochem. Physiol.* **19,** 45.

Saliot, A., and Barbier, M. (1973a). *Deep-Sea Res.* **20,** 1077.

Saliot, A., and Barbier, M. (1973b). *J. Exp. Mar. Biol. Ecol.* **13,** 207.

Scheuer, P. J. (1973). "Chemistry of Marine Natural Products," p. 201. Academic Press, New York.

Schmitz, F. J. (1978). *In* "Marine Natural Products," Vol. I (P. J. Scheuer, ed.), p. 245. Academic Press, New York.

Sheikh, Y. M., and Djerassi, C. (1975). *Steroids* **26,** 129.

Sica, D., and Zollo, F. (1978). *Tetrahedron Lett.* 837.

Sims, J. J., Donnell, M. S., Leary, J. V., and Lacy, G. H. (1975). *Antimicrob. Agents Chemother.* **7,** 320.

Singer, S. J., and Nicholson, G. L. (1972). *Science* **175,** 720.

Sivick, M., Marderossian, A. der, Ullucci, P., and Fenical, W. (1974). *In* "Food-Drugs from the Sea" (H. H. Webber and G. D. Ruggieri, eds.), p. 145. Puerto Rico.

Sondheimer, E., and Simeone, J. B. (1970). "Chemical Ecology," p. 336. Academic Press, New York.

Teshima, S. (1972). *Mem. Fac. Fish. Kagoshima Univ.* **21,** 69.

Toffler, A. (1974). "Le choc du futur," p. 180. Denoel, Paris.

Tusseau, D., Barbier, M., and Saliot, A. (1978). "Mission Orgon," Vol. II, Géochimie des Sédiments Marins Profonds, p. 253. CNRS, Paris.

Ui, J. (1971). *Rev. Int. Oceanog. Med. Fr.* **22,** 79.

Varkony, T. H., Smith, D. H., and Djerassi, C. (1978). *Tetrahedron* **34,** 841.

Viala, J., Devys, M., and Barbier, M. (1972). *Bull. Soc. Chim. Fr.* 3626.

Villanueva, V. R., Barbier, M., and Lederer, E. (1964). *Bull. Soc. Chim. Fr.* 1423.

Waksman, S. A. (1940). *In* "Les Antibiotiques" (J. Asselineau and J. P. Zalta, eds.), p. 364. Hermann, Paris.

Webber, H. H., and Ruggieri, G. D. (1974). *Proc. Food Drugs from the Sea, Puerto Rico* 509.

Westöö, G. (1966). *Acta Chem. Scand.* **20,** 2131.

Whittaker, R. H., and Feeny, P. P. (1971). *Science* **171,** 757.

Withers, N. C., Kokke, W. C. M. C., Rohmer, M., Fenical, W. H., and Djerassi, C. (1979). *Tetrahedron Lett.* 3605.

Woodwell, G. M., Craigh, P. P., and Johnson, H. A. (1971). *Science* **174,** 1101.

Wurster, S. F. (1968). *Science* **159,** 1474.

ZoBell, C. E. (1964). *Adv. Water Pollut. Res.* **3,** 85.

Appendix

Register of Known Compounds

PAUL J. SCHEUER

The following listing has resulted from a circular to some 50 marine natural product researchers. If you find this list useful and would like to see this feature continued in future volumes of this series, please submit your comments and/or entries to the Editor.

The arrangement is alphabetical by compound name. Additional information is tabulated in the following order: molecular formula, biological origin, geographical origin, previous reference if warranted, and present authors' names. Addresses are listed separately at the end.

COMPOUNDS

Batyl alcohol, diesters
  $C_{21}H_{44}O_3$ (alcohol)
  Sarcophyton pauciplicatum
    (Coelenterata)
  Red Sea
  Y. Kashman, S. Carmely
Cembrene-C
  $C_{20}H_{32}$
  Alcyonium flaccidum (Coelenterata)
  Red Sea
  D. J. Vanderah *et al., J. Org. Chem.* **43,**
    1614 (1978)
  Y. Kashman, A. Groweiss
Cholesta-5,23-dien-3β,25-diol (liagosterol)
  $C_{27}H_{44}O_2$
  Patella vulgata (Mollusca)
  Brittany, France
  F. Collignon-Thiennot, J. P. Allais, M.
    Barbier
Cholestanol
  $C_{27}H_{48}O$
  Orina arcoferus (Porifera)

Hamilton Bank, Labrador, Canada
J. F. Kingston, E. Benson, B. Gregory,
  A. G. Fallis
Cholesterol
  $C_{27}H_{46}O$
  Orina arcoferus, Geodia megastrella
    (Porifera)
  Hamilton Bank, Labrador, Canada
  J. F. Kingston, E. Benson, B. Gregory,
    A. G. Fallis
  Laurencia spectabilis (Rhodophyta)
  Sitka, AK
  P. B. Reichardt
cis-22-Dehydrocholesterol (occelasterol)
  $C_{27}H_{44}O$
  Orina arcoferus, Geodia megastrella
    (Porifera)
  Hamilton Bank, Labrador, Canada
  J. F. Kingston, E. Benson, B. Gregory,
    A. G. Fallis
trans-22-Dehydrocholesterol
  $C_{27}H_{44}O$
  Orina arcoferus, Geodia megastrella
    (Porifera)

187

MARINE NATURAL PRODUCTS
Copyright © 1980 by Academic Press, Inc.
All rights of reproduction in any form reserved.
ISBN 0-12-624004-3

Hamilton Bank, Labrador, Canada
J. F. Kingston, E. Benson, B. Gregory,
    A. G. Fallis
16-Deoxysarcophine (sarcophytoxide)
    $C_{20}H_{30}O_2$
Sarcophyton pauciplicatum, Sinularia
    vrijmoethi (Coelenterata)
Red Sea
    Y. Kashman et al., Tetrahedron 30, 3615
    (1974)
    Y. Kashman, S. Carmely
2,3-Dibromo-4,5-dihydroxybenzyl methyl
    ether
    $C_8H_8Br_2O_3$
Pterosiphonia bipinatta (Rhodophyta)
Sitka, AK
    P. B. Reichardt
3,5-Dibromo-4-hydroxybenzyl alcohol
    $C_7H_6Br_2O_2$
Odonthalia floccosa (Rhodophyta)
Sitka, AK
    P. B. Reichardt
3,4-threo-7,7-Dibromomethyl-3-methyl-
    3,4,8-trichloro-1,5(E),7(E)-octatriene
    $C_{10}H_{11}Br_2Cl_3$
Plocamium cartilagineum (Rhodophyta)
Bay of Naples
    F. Imperato et al., Experientia 33, 1273
    (1977)
    G. Cimino, S. DeRosa, S. DeStefano, G.
    Sodano
24-Ethylcholesterol (clionasterol,
    β-sitosterol)
    $C_{29}H_{50}O$
Orina arcoferus, Geodia megastrella
    (Porifera)
Hamilton Bank, Labrador, Canada
J. F. Kingston, E. Benson, B. Gregory,
    A. G. Fallis
Zostera marina (Magnoliophyta)
Izembek Lagoon, AK
    P. B. Reichardt
Fucosterol
    $C_{29}H_{48}O$
Orina arcoferus, Geodia megastrella
    (Porifera)
Hamilton Bank, Labrador, Canada
J. F. Kingston, E. Benson, B. Gregory,
    A. G. Fallis
Germacrene-C
    $C_{15}H_{24}$

Lithophyton arboreum, Stereonephthya
    condabilensis, Sinularia polydactyla
    (Coelenterata)
Red Sea
    K. Morikawa and Y. Hiroso,
    Tetrahedron Lett. 1799 (1969)
    Y. Kashman, N. Naveh, M. Bodner
Heptadecane
    $C_{17}H_{36}$
Zostera marina (Magnoliophyta)
Izembek Lagoon, AK
    P. B. Reichardt
Heteronemin
    $C_{29}H_{44}O_6$
Unidentified Enewetak sponge No. 14
Enewetak atoll
    R. Kazlauskas et al., Tetrahedron Lett.
    2631 (1976)
    F. Gerard, J. Weinberger, C. Ireland,
    P. J. Scheuer
Histamine dihydroiodide
    $C_5H_9N_3 \cdot 2HI$
Laminaria japonica (Phaeophyta)
Japan
    T. Kosuge
24-Methylcholesta-5,22-dien-3β-ol
    (brassicasterol)
    $C_{28}H_{46}O$
Orina arcoferus, Geodia megastrella
Hamilton Bank, Labrador, Canada
J. F. Kingston, E. Benson, B. Gregory,
    A. G. Fallis
24-Methylcholesterol (dihydrobrassi-
    casterol, campesterol)
    $C_{28}H_{48}O$
Orina arcoferus, Geodia megastrella
    (Porifera)
Hamilton Bank, Labrador, Canada
J. F. Kingston, E. Benson, B. Gregory,
    A. G. Fallis
α-Methylene-β-alanine, methyl ester
    amides
    $C_{24-28}H_{45-53}NO_3$
Heteronema erecta (Inodes erecta)
    (Porifera)
Red Sea
    Y. Kashman et al., Tetrahedron 29, 3655
    (1973); M. B. Yunker and P. J.
    Scheuer, Tetrahedron Lett. 4651
    (1978)
    Y. Kashman, M. Zviely

α-Methylene-β-alanine, methyl ester,
α-ketoamides
$C_{19}H_{33}NO_4$, $C_{20}H_{35}NO_4$
Spongia officinalis (Porifera)
Tahiti
M. B. Yunker and P. J. Scheuer,
  Tetrahedron Lett. 4651 (1978)
P. Amade, L. Chevolot
24-Methylenecholesterol
$C_{28}H_{46}O$
Orina arcoferus, Geodia megastrella
  (Porifera)
Hamilton Bank, Labrador, Canada
J. F. Kingston, E. Benson, B. Gregory,
  A. G. Fallis
Microcionin-2
$C_{15}H_{22}O$
Dysidea fragilis
Kaneohe Bay, Oahu, Hawaii
G. Cimino *et al., Tetrahedron Lett.* 3723
  (1975)
G. Schulte, P. J. Scheuer
Microcionin-4
$C_{15}H_{22}O$
Dysidea fragilis
Kaneohe Bay, Oahu, Hawaii
G. Cimino *et al., Tetrahedron Lett.* 3723
  (1975)
G. Schulte, P. J. Scheuer
Nephthenol
$C_{20}H_{34}O$
Scleronephthya corymbosa, Lithophyton
  arboreum, Sarcophyton decaryi,
  Lobophyton pauciflorum (Coelenterata)
Red Sea
F. J. Schmitz *et al., J. Chem. Soc.,*
  Chem. Commun. 407 (1974); B.
  Tursch *et al., Bull Soc. Chim. Belg.* **84,**
  767 (1975)
Y. Kashman, N. Naveh, M. Rotem, S.
  Carmely
Nonadecane
$C_{19}H_{40}$
Zostera marina (Magnoliophyta)
Izembek Lagoon, AK
P. B. Reichardt
24-Norcholesta-5,22-dien-3β-ol
$C_{26}H_{42}O$
Orina arcoferus, Geodia megastrella
  (Porifera)
Hamilton Bank, Labrador, Canada

J. F. Kingston, E. Benson, B. Gregory,
  A. G. Fallis
Pentadecane
$C_{15}H_{32}$
Zostera marina (Magnoliophyta)
Izembek Lagoon, AK
P. B. Reichardt
1,20-Pregnadien-3-one
$C_{21}H_{30}O$
Scleronephthya corymbosa
  (Coelenterata)
Red Sea
J. F. Kingston *et al., J. Chem. Soc.,*
  Perkin Trans. I 2064 (1979)
Y. Kashman, M. Rotem
1,4,20-Pregnatrien-3-one
$C_{21}H_{28}O$
Scleronephthya corymbosa
  (Coelenterata)
Red Sea
J. F. Kingston *et al., J. Chem. Soc.,*
  Perkin Trans. I, 2064 (1979)
Y. Kashman, M. Rotem
Pregna-1,4,20-trien-3-one
$C_{21}H_{28}O$
Gersemia sp. (Coelenterata)
Bonne Bay, Newfoundland, Canada
J. F. Kingston *et al., Tetrahedron Lett.*
  1601 (1977); M. D. Higgs and D. J.
  Faulkner, *Steroids* **30,** 379 (1977)
J. F. Kingston, B. Gregory, A. G. Fallis
Puupehenone
$C_{21}H_{28}O_3$
Inodes eubamma (Porifera)
Tahiti
B. N. Ravi *et al., Pure Appl. Chem.* **51,**
  1893 (1979)
P. Amade, L. Chevolot
Scalaradial
$C_{27}H_{40}O_4$
Heteronema erecta (Porifera)
Red Sea
G. Cimino *et al., Experientia* **29,** 934
  (1973)
Y. Kashman, M. Zviely
Spiniferin-2
$C_{15}H_{16}O$
Hypselodoris daniellae (Mollusca)
Ala Wai, Oahu, Hawaii
G. Cimino *et al., Experientia* **34,** 1425
  (1978)

G. Schulte, P. J. Scheuer
Taurine
$C_2H_7NO_3S$
Anemonia sulcata (Coelenterata)
Mediterranean
L. Béress and R. Béress, *Kieler Meeresforsch.* **27,** 117 (1971)
L. Béress
Thiofurodysinin acetate
$C_{17}H_{22}O_2S$
Unidentified sponge (Palau No. 12)
Palau
R. Kazlauskas *et al., Tetrahedron Lett.* 4951 (1978)
F. Gerard, C. Ireland, P. J. Scheuer

ADDRESSES
M. Barbier
Institut de Chimie des Substances Naturelles
Gif-sur-Yvette, France
L. Béress
Institut für Meereskunde
2300 Kiel, West Germany

L. Chevolot
Centre Océanologique de Bretagne
B.P. 337
29273 Brest Cedex, France
A. G. Fallis
Memorial University of Newfoundland
St. John's, Canada A1B 3X7
Y. Kashman
Tel-Aviv University
Tel-Aviv, Israel
T. Kosuge
Shizuoka College of Pharmacy
Shizuoka-ken
422 Japan
P. B. Reichardt
University of Alaska
Fairbanks, AK 99701
P. J. Scheuer
University of Hawaii at Manoa
Honolulu, HI 96822
G. Sodano
Laboratorio per la Chimica di Molecole di Interesse Biologico del CNR
Arco Felice (NA), Italy

Index